Modeling and Differential Equations in Biology

PURE AND APPLIED MATHEMATICS

A Program of Monographs, Textbooks and Lecture Notes

Contributions to *Lecture Notes in Pure and Applied Mathematics* are reproduced by direct photography of the author's typewritten manuscript. Potential authors are advised to submit preliminary manuscripts for review purposes. After acceptance, the author is responsible for preparing the final manuscript in camera-ready form, suitable for direct reproduction. Marcel Dekker, Inc. will furnish instructions to authors and special typing paper. Sample pages are reviewed and returned with our suggestions to assure quality control and the most attractive rendering of your manuscript. The publisher will also be happy to supervise and assist in all stages of the preparation of your camera-ready manuscript.

LECTURE NOTES
IN PURE AND APPLIED MATHEMATICS

1. *N. Jacobson,* Exceptional Lie Algebras
2. *L.-Å. Lindahl and F. Poulsen,* Thin Sets in Harmonic Analysis
3. *I. Satake,* Classification Theory of Semi-Simple Algebraic Groups
4. *F. Hirzebruch, W. D. Newmann, and S. S. Koh,* Differentiable Manifolds and Quadratic Forms
5. *I. Chavel,* Riemannian Symmetric Spaces of Rank One
6. *R. B. Burckel,* Characterization of C(X) Among Its Subalgebras
7. *B. R. McDonald, A. R. Magid, and K. C. Smith,* Ring Theory: Proceedings of the Oklahoma Conference
8. *Y.-T. Siu,* Techniques of Extension of Analytic Objects
9. *S. R. Caradus, W. E. Pfaffenberger, and B. Yood,* Calkin Algebras and Algebras of Operators on Banach Spaces
10. *E. O. Roxin, P.-T. Liu, and R. L. Sternberg,* Differential Games and Control Theory
11. *M. Orzech and C. Small,* The Brauer Group of Commutative Rings
12. *S. Thomeier,* Topology and Its Applications
13. *J. M. López and K. A. Ross,* Sidon Sets
14. *W. W. Comfort and S. Negrepontis,* Continuous Pseudometrics
15. *K. McKennon and J. M. Robertson,* Locally Convex Spaces
16. *M. Carmeli and S. Malin,* Representations of the Rotation and Lorentz Groups: An Introduction
17. *G. B. Seligman,* Rational Methods in Lie Algebras
18. *D. G. de Figueiredo,* Functional Analysis: Proceedings of the Brazilian Mathematical Society Symposium
19. *L. Cesari, R. Kannan, and J. D. Schuur,* Nonlinear Functional Analysis and Differential Equations: Proceedings of the Michigan State University Conference
20. *J. J. Schäffer,* Geometry of Spheres in Normed Spaces
21. *K. Yano and M. Kon,* Anti-Invariant Submanifolds
22. *W. V. Vasconcelos,* The Rings of Dimension Two
23. *R. E. Chandler,* Hausdorff Compactifications
24. *S. P. Franklin and B. V. S. Thomas,* Topology: Proceedings of the Memphis State University Conference
25. *S. K. Jain,* Ring Theory: Proceedings of the Ohio University Conference
26. *B. R. McDonald and R. A. Morris,* Ring Theory II: Proceedings of the Second Oklahoma Conference
27. *R. B. Mura and A. Rhemtulla,* Orderable Groups
28. *J. R. Graef,* Stability of Dynamical Systems: Theory and Applications
29. *H.-C. Wang,* Homogeneous Banach Algebras
30. *E. O. Roxin, P.-T. Liu, and R. L. Sternberg,* Differential Games and Control Theory II
31. *R. D. Porter,* Introduction to Fibre Bundles
32. *M. Altman,* Contractors and Contractor Directions Theory and Applications
33. *J. S. Golan,* Decomposition and Dimension in Module Categories
34. *G. Fairweather,* Finite Element Galerkin Methods for Differential Equations
35. *J. D. Sally,* Numbers of Generators of Ideals in Local Rings
36. *S. S. Miller,* Complex Analysis: Proceedings of the S.U.N.Y. Brockport Conference
37. *R. Gordon,* Representation Theory of Algebras: Proceedings of the Philadelphia Conference
38. *M. Goto and F. D. Grosshans,* Semisimple Lie Algebras
39. *A. I. Arruda, N. C. A. da Costa, and R. Chuaqui,* Mathematical Logic: Proceedings of the First Brazilian Conference
40. *F. Van Oystaeyen,* Ring Theory: Proceedings of the 1977 Antwerp Conference

41. *F. Van Oystaeyen and A. Verschoren,* Reflectors and Localization: Application to Sheaf Theory
42. *M. Satyanarayana,* Positively Ordered Semigroups
43. *D. L. Russell,* Mathematics of Finite-Dimensional Control Systems
44. *P.-T. Liu and E. Roxin,* Differential Games and Control Theory III: Proceedings of the Third Kingston Conference, Part A
45. *A. Geramita and J. Seberry,* Orthogonal Designs: Quadratic Forms and Hadamard Matrices
46. *J. Cigler, V. Losert, and P. Michor,* Banach Modules and Functors on Categories of Banach Spaces
47. *P.-T. Liu and J. G. Sutinen,* Control Theory in Mathematical Economics: Proceedings of the Third Kingston Conference, Part B
48. *C. Byrnes,* Partial Differential Equations and Geometry
49. *G. Klambauer,* Problems and Propositions in Analysis
50. *J. Knopfmacher,* Analytic Arithmetic of Algebraic Function Fields
51. *F. Van Oystaeyen,* Ring Theory: Proceedings of the 1978 Antwerp Conference
52. *B. Kedem,* Binary Time Series
53. *J. Barros-Neto and R. A. Artino,* Hypoelliptic Boundary-Value Problems
54. *R. L. Sternberg, A. J. Kalinowski, and J. S. Papadakis,* Nonlinear Partial Differential Equations in Engineering and Applied Science
55. *B. R. McDonald,* Ring Theory and Algebra III: Proceedings of The Third Oklahoma Conference
56. *J. S. Golan,* Structure Sheaves over a Noncommutative Ring
57. *T. V. Narayana, R. M. Mathsen, and J. G. Williams,* Combinatorics, Representation Theory and Statistical Methods in Groups: YOUNG DAY Proceedings
58. *T. A. Burton,* Modeling and Differential Equations in Biology

Other Volumes in Preparation

Modeling
and Differential Equations
in Biology

edited by

T. A. BURTON

Department of Mathematics
Southern Illinois University
Carbondale, Illinois

MARCEL DEKKER, INC. New York and Basel

Library of Congress Cataloging in Publication Data

Main entry under title:

Modeling and differential equations in biology.

(Lecture notes in pure and applied mathematics ; 58)
Papers presented at a regional conference held at Southern Illinois University, Carbondale, Ill., June 5-9, 1978.
1. Biology--Mathematical models--Congresses. 2. Differential equations--Congresses. I. Burton, Theodore Allen.
QH323.5.M62 574'.0724 80-19894
ISBN 0-8247-1075-4

MARCEL DEKKER, INC.

270 Madison Avenue, New York, New York 10016

Current printing (last digit):

10 9 8 7 6 5 4 3 2 1

PRINTED IN THE UNITED STATES OF AMERICA

PREFACE

This volume contains the invited lectures and contributed papers presented at the regional conference on Modeling and Differential Equations in Biology, supported by the Conference Board of the Mathematical Sciences and the National Science Foundation, which was held at Southern Illinois University in Carbondale, Illinois from June 5 to 9, 1978. Professor H. T. Banks of Brown University was the Principal Lecturer. His lecture notes will appear in a separate volume published for the Conference Board.

Invited lectures were given by Professors T. Hallam, H. Thames, Jr., and P. Waltman. Their lectures, together with substantial supporting material, are contained in this volume. During the conference, Professor Steve Merrill was invited to contribute an expository paper on immune response. That is the only paper included here which was not presented at the conference.

Professors N. Alikakos, G. Bécus, G. Butler, W. Fair, B. Goh, A. Hastings, and S. Merrill contributed one or more papers during the week and these are all included (some have co-authors).

Each of the contributed papers was subjected to a refereeing process similar to that for a journal article.

The conference itself was an exceptionally exciting experience. Most of the participants were mathematicians specializing in differential equations. The lectures of Banks and the other speakers opened up broad areas of application to the participants. Each author was asked to provide substantial expository material in his article. It is our hope that this volume will do much to stimulate research in the direction of applications of differential equations to biology.

I wish to thank each of the speakers, the Conference Board, and the National Science Foundation. In particular, I am indebted to Ms. Sharon Champion for her assistance throughout the planning of the conference, during the conference itself, and for the typing of the manuscripts for the proceedings.

Funds for the typing of the manuscripts were provided by the Department of Mathematics and the College of Liberal Arts of Southern Illinois University at Carbondale, Illinois.

T. A. Burton

CONTENTS

CONTRIBUTORS

N. ALIKAKOS *Department of Mathematics, Purdue University, West Lafayette, Indiana*

GEORGES A. BÉCUS *Engineering Science Department, University of Cincinnati, Cincinnati, Ohio*

G. J. BUTLER *Department of Mathematics, University of Alberta, Edmonton, Alberta, Canada*

WYMAN FAIR *Department of Mathematical Sciences, Drexel University, Philadelphia, Pennsylvania*

B. S. GOH *Mathematics Department, University of Western Australia, Nedlands, Australia*

THOMAS G. HALLAM *Department of Mathematics and Program in Ecology, University of Tennessee, Knoxville, Tennessee*

ALAN HASTINGS* *Department of Mathematics, University of California, Davis, California*

SZE-BI HSU† *Department of Applied Mathematics, National Chiao Tung University, Hsinchu, Taiwan*

STEPHEN P. HUBBELL *Department of Zoology, University of Iowa, Iowa City, Iowa*

JESSE A. LOGAN *The National Resource Ecology Laboratory and Department of Zoology and Entomology, Colorado State University, Fort Collins, Colorado*

STEPHEN J. MERRILL *Department of Mathematics and Statistics, Marquette University, Milwaukee, Wisconsin*

Current affiliations:
*Department of Mathematics, University of California, Davis, California

†Department of Applied Mathematics, National Chiao Tung University, Hsinchu, Taiwan

CHRIS RORRES *Department of Mathematics, Drexel University, Philadelphia, Pennsylvania*

HOWARD D. THAMES, JR. *The University of Texas System Cancer Center, M. D. Anderson Hospital and Tumor Institute, Department of Biomathematics, Houston, Texas*

PAUL WALTMAN* *Department of Mathematics, University of Iowa, Iowa City, Iowa*

DAVID J. WOLLKIND *Department of Pure and Applied Mathematics, Washington State University, Pullman, Washington*

**Current affiliation:* Department of Mathematics, University of Iowa, Iowa City, Iowa

Modeling and Differential Equations in Biology

PERSISTENCE IN LOTKA–VOLTERRA MODELS OF FOOD CHAINS AND COMPETITION

Thomas G. Hallam

Department of Mathematics and Program in Ecology
University of Tennessee
Knoxville, Tennessee

INTRODUCTION

In this article, I will survey some recent contributions to the persistence-extinction theory of mathematical models of ecosystems with emphasis upon a class of models with linear per capita growth rates. Initially investigated by A. J. Lotka and V. Volterra over fifty years ago, this class has long been an important cornerstone of mathematical ecology (although from an ecological perspective it is certainly deficient).

The obvious importance of extinction phenomena is magnified by its frequency of occurrence. Levins [12] has estimated that there have been 10^8 to 10^9 species extinctions since the onset of the Cambrian era some 10^8 to 10^9 years ago. These phenomena are also of current interest as illustrated by recent public concern for endangered species. Such actions are important as the consequences of extinction are not yet clearly understood although certain trends have been observed. For example, a top predator in an ecosystem can have an important effect upon structure of the food web. Extinction of such species has resulted in an unexpected collapse of the web. These predation effects have been observed by Paine [18] for an aquatic system and by Connell [2] for terrestrial ecosystems. Unfortunately, mathematical analyses have not yet been developed to completely explain these general characteristics. The work described here indicates that present model knowledge is only a good start on very complex problems.

The models considered here are of the form

$$\frac{dx_i}{dt} = x_i F_i(x_1, x_2, \ldots, x_n), \qquad i = 1, 2, \ldots, n \tag{1}$$

Each F_i is a continuous function from R_+^n, the nonnegative cone in R^n, to R and is sufficiently smooth to guarantee that initial value problems associated with (1) have unique solutions in the population orthant, R_+^n. *Extinction* occurs in (1) if there is a solution $\phi = (\phi_1, \phi_2, \ldots, \phi_n)$ of (1) with $\phi_i(0) > 0$, $i = 1, 2, \ldots, n$, and having a component ϕ_j which satisfies $\lim_{t \to \tau} \phi_j(t) = 0$ for some τ in $(0, \infty]$. *Persistence* is the negation of extinction and requires that each solution ϕ of (1) with $\phi_i(0) > 0$, $i = 1, 2, \ldots, n$ satisfy $\limsup_{t \to \tau} \phi_i(t) > 0$ for all $\tau \in (0, T_\phi]$ and all $i = 1, 2, \ldots, n$ wherein $[0, T_\phi]$ denotes the maximal interval of existence of ϕ.

Extinction is a local property while persistence is a global concept. Much effort in studies of ecological models has been directed to local stability analyses and numerical developments (e.g., [19, 4, 15, 7, 22]). To investigate a global phenomenon such as persistence analytical studies are required.

The sections of this article are structured by the type of the food web considered. The second section discusses some recent results on persistence of simple food chains. Persistence in Lotka-Volterra food chains can be determined through knowledge of a single parameter. This parameter is a function of interspecific and intraspecific interaction coefficients as well as the length of the food chain.

When trophic levels are more complex due to the presence of additional species, the number of parameters required to determine persistence increases. This property may be observed in a two dimensional Lotka-Volterra competition model

$$\frac{dx}{dt} = x(a - bx - cy),$$
$$\frac{dy}{dt} = y(e - fx - gy) \qquad (2)$$

where a, b, c, e, f, and g are positive parameters representing a fixed constant ecology. The number of persistence parameters is two; namely, $\rho = be - af$ and $\sigma = ag - ce$. Both ρ and σ need to be positive in order that the model represents a persistent community. The third section demonstrates that the effects of complexity on persistence of a community composed of three competing species are even more complicated than might be anticipated from the two dimensional model.

PERSISTENCE IN FOOD CHAINS

A simple food chain has trophic levels which are functionally regarded as a single species. The predator-prey relationship between adjoining components of simple food chains has the dynamics of each trophic level species governed by the species occupying the adjoining trophic levels.

My first remarks are limited to Lotka-Volterra models of simple food chains; specifically, we consider the model

$$\begin{aligned}
\frac{dx_1}{dt} &= x_1(a_{10} - a_{11} x_1 - a_{12} x_2) \\
\frac{dx_2}{dt} &= x_2(-a_{20} + a_{21} x_1 - a_{23} x_3) \\
&\vdots \\
\frac{dx_{n-1}}{dt} &= x_{n-1}(-a_{n-1,0} + a_{n-1,n-2}x_{n-2} - a_{n-1,n} x_n) \\
\frac{dx_n}{dt} &= x_n(-a_{n0} + a_{n,n-1} x_{n-1})
\end{aligned} \qquad (3)$$

where all coefficients a_{ij} are positive constants with the exception of a_{11} which is nonnegative. The constant a_{11} determines the behavior of the resource (lowest trophic) level x_1 in the absence of additional components of the chain. When $a_{11} = 0$ the resource supply is growing exponentially while $a_{11} > 0$ indicates there is density dependent regulation. The existence of a resource carrying capacity has important implications for the asymptotic behavior of system (3). Assuming a carrying capacity, solutions initially in the population orthant are bounded. This result was indicated by Freedman and Waltman [5] for a three dimensional food chain and by Gard and Hallam [6] in an n-dimensional setting. If the resource is allowed to grow exponentially in the absence of other species, the analogous result utilized by Gard and Hallam [6] was that all solutions with a component that goes extinct are bounded. As we shall see later, this conclusion is about the best possible as solutions of (3) with $a_{11} = 0$ are not, in general, bounded.

To state our first result on persistence-extinction, some notation is required. Define the parameter μ which will determine persistence and extinction by

$$\mu = a_{10} - \left(\frac{a_{11}}{a_{21}}\right)\left[a_{20} + \sum_{j=2}^{k}\left(\prod_{i=2}^{j}\frac{a_{2i-2,\ 2i-1}}{a_{2i,\ 2i-1}}\right)a_{2j,\ 0}\right] - \sum_{j=1}^{\ell}\left(\prod_{i=1}^{j}\frac{a_{2i-1,\ 2i}}{a_{2i+1,\ 2i}}\right)a_{2j+1,\ 0}$$

where

$$K = \begin{cases} \dfrac{n}{2} & \text{if } n \text{ is even} \\[2ex] \dfrac{n-1}{2} & \text{if } n \text{ is odd} \end{cases}$$

and

$$\ell = \begin{cases} \dfrac{n}{2} - 1 & \text{if } n \text{ is even} \\[2ex] \dfrac{n-1}{2} & \text{if } n \text{ is odd} \end{cases}$$

THEOREM 1 Suppose the resource level in (3) has a carrying capacity ($a_{10}/a_{11} > 0$). Then, if μ is positive, system (3) is persistent. If μ is negative then system (3) is not persistent.

Outline of Proof. I will just outline the main steps in the proofs; the details may be found in Gard and Hallam [6]. Several properties of dynamical systems play important roles in the proof.

1. The bounding hyperplanes are invariant manifolds for system (3).
2. Any extinction trajectory converges as t approaches infinity to a set containing recurrent solutions.
3. Any recurrent solution in a bounding hyperplane and in the ω-limit set of an extinction trajectory is contained in a minimal bounding hyperplane.

For appropriate choices of r_i the function $\rho = \Pi_{i=1}^{n} x_i^{r_i}$ is increasing along trajectories of (3) that are close to this minimal bounding hyperplane provided $\mu > 0$. These facts can be employed to demonstrate persistence. The extinction result when $\mu < 0$ also utilizes a similar system transformation and differential inequalities.

When the resource level is not regulated by density dependent factors, a classification of persistence can be obtained in terms of the parameter

$$\mu_0 = a_{10} - \sum_{j=1}^{m} a_{2j+1,\ 0} \prod_{i=1}^{j} \frac{a_{2i-1,\ 2i}}{a_{2i+1,\ 2i}}$$

wherein m is defined by n = 2m + 1 or n = 2m + 2 depending upon whether n is odd or even respectively. Extinction results if $\mu_0 < 0$ while persistence ensues whenever $\mu_0 > 0$. Odd dimensional models of form (3) with $a_{11} = 0$ have an interesting behavior in that persistence can result even though there is no equilibrium in the interior of the population orthant.

Coste, Peyrand, and Coullet [3] have shown that for bounded Lotka-Volterra systems, persistence implies the existence of an interior equilibrium in the population orthant. Thus, for odd dimensional, simple food chains of Lotka-Volterra type, there must exist solutions which are unbounded.

This situation contrasts with model (3) when a carrying capacity exists. Persistence is equivalent to the existence of a positive, asymptotically stable equilibrium which is a global attractor. Harrison [10] has demonstrated that whenever (3) has an interior equilibrium in the population orthant, it is a global attractor. The persistence condition $\mu > 0$

is equivalent to the existence of an interior equilibrium, so persistence in simple Lotka-Volterra food chains may be viewed dynamically in terms of a global attractor in the interior of the population orthant. Goh [8] has studied related Lotka-Volterra systems that are also globally asymptotically stable.

Related work on food chains includes that of Freedman and Waltman [5] who consider Lotka-Volterra and other more general food chain models. Saunders and Bazin [20] also study food chains with general response formulations. Their techniques are primarily those of locally stability analysis.

PERSISTENCE IN LOTKA-VOLTERRA MODELS OF COMPETITION

In this section, I shall discuss some effects of trophic level complexity upon persistence of the ecosystem. The underlying problem of stability vs. complexity has been widely debated [15, 21] without resolution. The results below do little to assist in this dilemma from the perspective that the conclusions do not appear to be sufficiently unified to facilitate general understanding. However, they illustrate well the difficulties which arise in structuring a general theory.

Studies of numerous documented food webs [1, 11] show that food chains are relatively short, usually composed of three to four species with the longest documented being one of length six. From the complexity aspect, there seems to be little that can be said concerning the number of species that can occupy a given trophic level. A study of the Lotka-Volterra competition model

$$\begin{aligned} \frac{dx}{dt} &= x(a - bx - cy - dz) \\ \frac{dy}{dt} &= y(e - fx - gy - hz) \\ \frac{dz}{dt} &= z(i - jx - ky - \ell z) \end{aligned} \tag{4}$$

is an initial attempt to understand why species on the same trophic level interact in the ways that they do. The coefficients in (4) are assumed to be positive constants.

Asymptotic behavior of (4) has been studied by Rescigno [19], Gilpin [7], May and Leonard [15], and Strobeck [22]. These articles generally employ classical stability approaches or numerical procedures to discuss behavior aspects of competitive systems. The next results complement those above in that they are global in character, and are due to Hallam, Svoboda, and Gard [9]. The feasible types of extinction that can occur for Lotka-Volterra competitive systems are delineated first, then persistent arrangements that can be determined from knowledge of two species interactions are given.

The next result indicates that complete extinction cannot occur in (4).

THEOREM 2 For any solution $(x(t), y(t), z(t))$ of (4) with $x(0) > 0$, $y(0) > 0$, $z(0) > 0$, the limits: $\lim_{t\to\infty} x(t) = 0$; $\lim_{t\to\infty} y(t) = 0$; $\lim_{t\to\infty} z(t) = 0$; cannot occur simultaneously.

This result is trivial by the geometric configuration of (4) about the origin. However an analytic proof will be indicated to give the flavor of the persistence-extinction function technique which is utilized in the main stated results on food chains and competition. Consider the auxiliary function $\rho = x\,y\,z$ and denote by $\dot{\rho}$ the total time derivative of ρ along trajectories of (4). The function $\dot{\rho}$ must satisfy $\dot{\rho} = \rho[a + e + i - x(b + f + j) - y(c + g + k) - z(d + h + \ell)]$. When total extinction occurs, then there is a T such that whenever $t \geq T$,

$$x(t) < \frac{a + e + i}{4(b + f + j)}$$

$$y(t) < \frac{a + e + i}{4(c + g + k)}$$

$$z(t) < \frac{a + e + i}{4(d + h + \ell)}$$

It follows for $\eta = (a + e + i)/4$, that

$$\dot{\rho} \geq \rho\eta, \quad t \geq T$$

and, hence, that $\rho(t) \geq \rho(T)\, e^{\eta(t-T)}$, $t \geq T$. This implies that $\lim_{t\to\infty} \rho(t) = \infty$ and contradicts the fact that simultaneous extinction of x, y, and z implies $\lim_{t\to\infty} \rho(t) = 0$. This completes the indication of the analytical proof.

Consequently we note that the main extinction results for Lotka-Volterra competitive systems are for extinctions of one or two species.

Classifications of these types of extinctions can be phrased in terms of the model parameters as is indicated by the following two species extinction result.

THEOREM 3 If there exists a solution $(x(t), y(t), z(t))$ of (4) with the asymptotic behavior $\lim_{t\to\infty} x(t) = 0$, $\lim_{t\to\infty} y(t) = 0$, and $\lim_{t\to\infty} \sup z(t) > 0$, then $di > a\ell$ and $hi > e\ell$. Conversely, if $hi > e\ell$ and $di > a\ell$ then there exists a nonempty region, R_1, contained in the population octant where each trajectory of (4) with $(x(0), y(0), z(0)) \in R_1$ satisfies $\lim_{t\to\infty} x(t) = \lim_{t\to\infty} y(t) = 0$ and $\lim_{t\to\infty} \sup z(t) > 0$.

For the details of the proof the reader is referred to Hallam et al. [9]. The general idea is to employ a suitable persistence function with differential inequality theory.

Single species extinction conditions are the content of the next result.

THEOREM 4 If there exists a solution $(x(t), y(t), z(t))$ of (4) with the behavior $\lim\sup_{t\to\infty} x(t) > 0$, $\lim\sup_{t\to\infty} y(t) > 0$, and $\lim\sup_{t\to\infty} z(t) = 0$ then $(ag - ce)(be - af) \geqq 0$ and $\gamma_z = i - j(\frac{ag - ce}{bg - cf}) - k(\frac{be - af}{bg - cf}) \leqq 0$.

This result may be verified by using a qualitative approach or by a persistence function approach [9]. The partial converse, similar to the last statement in Theorem 3, in the robust case wherein $ag > ce$, $be > af$ and $\gamma_z < 0$ follows directly by use of linearization techniques. With the extinction results delineated, persistence may be ascertained by the brute force method of eliminating all extinction possibilities.

An objective of mathematical modeling is to utilize hypotheses and obtain conclusions that are realistic and feasible. Information about the competitive interactions between two species is sometimes available (e.g., [13, 17, 23]). Hence, we study persistence in three species models by employing two species interactions as a basic hypothesis.

To facilitate stating the results, the three possible outcomes of two dimensional Lotka-Volterra competition (see (2)) are given designations. If species x always dominates y, we use the symbol $x \gg y$. If species x and y coexist in a stable manner, the symbol $x \leftrightarrow y$ is used. If the winner of the competition depends upon the initial position of the populations, then this outcome is denoted by $x \overset{\uparrow}{\underset{\downarrow}{\leftrightarrow}} y$. All outcomes are related to the existence and the stability of an equilibrium in the population quadrant. When (2) does not have an interior equilibrium, $x \gg y$ or $y \gg x$ results. The outcome $x \leftrightarrow y$ is equivalent to the existence of a stable equilibrium in the interior of the population quadrant while $x \overset{\uparrow}{\underset{\downarrow}{\leftrightarrow}} y$ corresponds to an unstable equilibrium.

There are four arrangements, disregarding relabeled species, of the three combinations of two species interactions that can lead to persistence; these are

A.1. $x \gg y$, $z \gg x$, $y \gg z$

A.2. $x \leftrightarrow y$, $x \gg z$, $z \gg y$

A.3. $x \leftrightarrow y$, $x \leftrightarrow z$, $y \gg z$

A.4. $x \leftrightarrow y$, $x \leftrightarrow z$, $z \leftrightarrow y$

The nontransitive arrangement A.1. [15, 7] is unique in that persistence is implicit in the arrangement. All other arrangements require that additional criteria, measures of the ability of a population to invade an established equilibrium community, hold. For example, in arrangement A.2., the ability of species z to invade the xy-community is measured by the parameter γ_z of Theorem 4. If γ_z is positive, invasion by z is successful and the system persists. If an arrangement contains any pair of species that interact in stable competition, the complementary species must exhibit a positive invasion capability when the stable pair is at equilibrium density.

All cases of persistence and extinction in three dimensional Lotka-Volterra competition are examined in Hallam et al. [9]. There are some model conclusions that merit mention. Any pair of species that interacts in an unstable manner cannot be a component of a persistent three dimensional Lotka-Volterra competitive system. This phenomenon has also been observed in Lotka-Volterra models of a predatory or mutualistic species invading a competitive community. Extinction in these systems can occur in one, two, or three different regions. These regions need not correspond to extinction of the same species.

SUMMARY

Persistence-extinction in Lotka-Volterra models of simple food chains is well understood. A single parameter suffices to explain the phenomenon. For such models with a resource carrying capacity, persistence is equivalent to the existence of an equilibrium in the interior of the population orthant. This equilibrium is always globally asymptotically stable. If the resource population is assumed to grow exponentially, persistence can occur when there does not exist an interior equilibrium and there exist unbounded solutions.

In three dimensional Lotka-Volterra models of competition, the effects of complexity on persistence become pronounced. Knowledge of several parameters is now required to determine persistence. When two species interactions are assumed known, persistence can be determined by an elimination of the cases where extinction occurs. A single arrangement, the nontransitive set $x \gg y$, $y \gg z$, $z \gg x$, yields persistence implicitly. All remaining persistent arrangements contain a stable coexisting pair of competitors. The persistence conditions require that the species complementary to the stable competitive subcommunity must be able to mount a successful invasion.

ACKNOWLEDGMENT

This research was supported in part by the National Science Foundation under Interagency Agreement 40-700-78 with the U.S. Department of Energy, Oak Ridge National Laboratory (operated by Union Carbide Corporation under contract W-7405-eng-26 with the U.S. Department of Energy).

REFERENCES

1. J. E. Cohen, *Food Webs and Niche Space*, Princeton Univ. Press, Princeton, (1978).

2. J. H. Connell, "Some mechanisms producing structure in natural communities, *Ecology and Evolution of Communities*, M.S. Cody and J. M. Diamond (Eds.) (1975), 460-491.

3. J. Coste, J. Peyrand and P. Coullet, "Does complexity favor existence of persistent systems?" *J. Theor. Biol.* 73(1978), 359-362.

4. D. L. DeAngelis, "Stability and connectance in food web models," *Ecology* 56(1975), 238-243.

5. H. I. Freedman and P. Waltman, "Mathematical analysis of some three species food-chain models," *Math. Biosciences* 33(1977), 257-276.

6. T. C. Gard and T. G. Hallam, "Persistence in food webs: I. Lotka-Volterra food chains," *Bull. Math. Biol.* 41(1979), 877-891.

7. M. E. Gilpin, "Limit cycles in competitive communities," *Amer. Natur.* 109(1975), 51-60.

8. B. S. Goh, "Global stability in many-species systems," *Amer. Natur.* 111(1977), 135-143.

9. T. G. Hallam, L. J. Svoboda and T. C. Gard, "Persistence and extinction in three species Lotka-Volterra competition," *Math. Biosciences* 46(1979), 117-124.

10. G. W. Harrison, "Global stability of food chains," *Amer. Natur.* (1978).

11. J. H. Lawton and S. L. Pimm, "Population dynamics and the length of food chains," *Nature* 272(1978), 190.

12. R. Levins, "Extinction," *Some Mathematical Questions in Biology.* American Math. Soc. 2(1970), 75-107.

13. E. J. Maly, "Interactions among the predatory rotifer *Asplanchna* and two prey, *Paramecium* and *Euglena*," *Ecology* 56(1975), 346-358.

14. R. M. May, *Stability and Complexity in Model Ecosystems.* 2nd ed. Princeton Univ. Press, Princeton, N. J. (1974).

15. R. M. May and W. J. Leonard, "Nonlinear aspects of competition between three species," *SIAM J. Appl. Math.* 129(1975), 243-253.

16. W. W. Murdoch and A. Oaten, "Predation and population stability," *Adv. Ecol. Res.* 9(1975), 2-131.

17. W. E. Neill, "Experimental studies of microcrustacean competition, community composition, and efficiency of resource utilization," *Ecology* 56(1975), 809-926.

18. R. T. Paine, "Food web complexity and species diversity," *Amer. Natur.* 100(1966), 65-75.

19. A. Rescigno, "The struggle for life: II. three competitors," *Bull. Math. Biophys.* 30(1968), 291-298.

20. P. T. Saunders and M. J. Bazin, "On stability of food chains," *J. Theor. Biol.* 52(1975), 121-142 .

21. D. D. Siljak, "Connective stability of complex systems," *Nature* 249 (1974), 280.

22. C. Strobeck, "N-species competition," *Ecology* 54(1978), 650-654.

23. H. M. Wilbur, "Competition, predation, and the structure of the *Ambystoma-rana sylvatica* community," *Ecology* 53(1973), 3-21.

MATHEMATICAL MODELS OF HUMORAL IMMUNE RESPONSE

Stephen J. Merrill

Department of Mathematics and Statistics
Marquette University
Milwaukee, Wisconsin

INTRODUCTION

This is a survey of several mathematical models of the humoral immune response. They are predominantly models of the population levels and interaction of various cell types involved in the response as well as the associated production of antibody and the removal of antigen from the system.

Each model is discussed separately and the notation is as it appears in the original papers. Some simplifications in the models have been made to make comparison with the others easier. These are generally noted in the discussion.

The models here all involve differential equations. The mathematical analysis of the systems as it has been developed is given in capsulary form, both to indicate the methods and maybe suggest extensions to these results.

OUTLINE OF THE IMMUNE RESPONSE

The immune system is a set of organs, cells, and proteins which responds to the presence of a foreign substance in the body. The response is twofold: (1) the *humoral response* and (2) the *cell-mediated response*. The humoral response results in the production of a heterogeneous collection of proteins called antibodies which are specifically aimed at a molecular pattern or antigenic determinant on the surface of the foreign substance called an antigen. These antibodies react and bind with the antigen forming antigen-antibody complexes which hasten the destruction and removal of the antigen from the body. The cell-mediated response involves a large number of different phenomena which result in destruction of the antigen without presence of antibody. This response is especially important in the response against fungi, certain viruses, and protozoa.

Both responses have small lymphocytes as important participants. These come in two types: (1) the *B-cell* which will evolve into antibody-producing cells, and (2) the *T-cell* which is responsible for the cell-mediated response and which is necessary for a strong humoral response against most antigens.

The surface of a B-cell is covered by approximately 100,000 molecules of antibody which it uses as *receptors* for a particular antigenic determinant. However, most antigens require some "processing" by other cells, especially T-cells and macrophages before they can stimulate a B-cell. Many aspects of this interaction between cells and cell products from T-cells are not yet understood. (A complete survey of known effects of T-cells on the humoral response can be found in [28] or [11].) When a receptor makes contact with (properly processed) antigen with that antigenic determinant, the receptor binds the antigen to the surface of the B-cell. If little of the antigen is present and few of the receptors bind antigen, no change in the B-cell is observed (this is called low dose unresponsiveness). When the number of these bindings is large enough, the B-cell is "stimulated." Stimulation results in blast transformation which involves an increase in the size and the number of the internal structures needed for protein synthesis and the distribution of the product (antibody) to the surface. This stage is also noted by increases in the rates of macromolecular synthesis and mitosis. For example, the unstimulated B-cell divides approximately once in five years while the large blast cells double in 6-48 hours. The resulting mass of cells is called a *clone*. In this way, the number of lymphocytes able to recognize a particular antigen is amplified.

When the clone is large enough, determined by the amount of antigen present, the cells begin to differentiate into *plasma cells* which are short-lived cells, principally responsible for the production of antibody, and *memory cells*, small long-lived cells which are responsible for a quick strong response should that same antigen be encountered again. When the differentiation is complete, the plasma cells begin to produce antibody whose binding sites are identical to those of the receptors on the original B-cell. This antibody is released from the plasma cell into the circulatory system where it binds with antigen and hastens its elimination by other mechanisms. The plasma cell lives for 10-14 days and then dies without division. The memory cell eventually shrinks to the size of the small B-cell and is able to respond to antigen after essentially all the free antibody has disappeared from the system. This response is called the *primary response* if that antigen has not previously elicited a response, and the *secondary response* if it has.

In comparison with the primary response, the secondary response is characterized by a lower threshold dose of antigen, a shorter lag between introduction of antigen and production of antibody, a higher rate and longer persistence of antibody synthesis and higher peak concentrations of antibody. The secondary response is responsible for what is generally known as "immunity."

An antigen present in sufficient concentration may fail to elicit a humoral response if no B-cells which recognizes that particular antigen are present or able to respond. This condition is called *tolerance* if this failure is specific for the antigen. Tolerance can be induced in two ways: long-term presence of small amounts of the antigen (*low zone tolerance*) and long-term high-level presence (*high zone tolerance*). Tolerance makes it possible for lymphocytes to avoid making an immune response against substances naturally found in the body. See [36] for a survey of immunologic tolerance.

A general description of the immune response may be found in [10].

HISTORY OF MATHEMATICAL MODELLING OF HUMORAL IMMUNE RESPONSE

Many researchers have proposed models of certain aspects of the humoral immune response. In 1966, Hege and Cole [16] constructed a model relating the changes in circulating antibody with the numbers of antibody-producing cells. Jilek [22, 23] and Jilek and Sterzl [24, 25] developed stochastic models for the dynamics of the response in terms of evaluating the probability of antigen meeting an immunocompetent B-cell. Cohen [8, 9] in 1970 and 1971 proposed a model of the response, assuming the response of the B-cell depends on the ratio of antigen molecules bound to two adjacent receptors on the B-cell, to study stimulation and tolerance.

Bell [1] in 1970 and 1971, using as a main hypothesis the clonal selection theory of Burnet [7], proposed a model to study the humoral response to a challenge of injected antigen which had no replicating ability. The system of differential equations resulting was solved numerically and the results were compared to the known response when 2, 4 dinitrophenyl-bovine γ-globulin (DNP-BGG) was injected into rabbits. The model consisted of six differential equations which governed the numbers of B-cells able to respond to a particular antigen, proliferating cells, plasma, cells, memory cells, and concentrations of antibody and antigen. In the later papers, he studied the response of a finite number of lymphocytes with different affinities for the antigen, allowing the antigen to have more than one combining site on the molecule (multivalent) and possible mechanisms for tolerance. A mathematical drawback of his model is that the size of the system of differential equations makes study difficult and information on the qualitative properties of the model can only be surmised from numerical examples.

To obtain a simplest possible model of the immune response to a replicating antigen, Bell [2] posed a model in which he viewed the antibody-antigen interaction as a predator-prey situation. This model was small enough (only two differential equations) to allow qualitative study which resulted in identification of important features of the model: places of equilibrium or *critical points* where antibody-antigen competition results in no change in either concentration, *periodic solutions* where antibody-antigen concentrations change in a predictable and recurrent manner, and *asymptotic behavior* which gives conditions where antigen concentration goes to zero or where antigen concentration becomes too large and the host dies. This qualitative study was done in [2] by Bell and in [39] by Pimbley. This model also exhibited behavior which was impossible for a living system (oscillations of ever-increasing amplitude).

In 1974, Bell proposed a second "predator-prey" model in a paper by Pimbley [40], which added one differential equation governing the growth of the B-cell population to the two in the earlier model. This equation acts to change the "environment" in which the antibody-antigen interactions are occurring. The qualitative features of this model were studied by Pimbley [40-42] and by Hsü and Kazarinoff [19]. Pimbley was able to identify critical (bifurcation) values of the parameters so that around these critical values, the qualitative behavior of the model would change and in this way identified one-parameter families of periodic solutions using the Hopf bifurcation theorem.

A different direction has been taken by Richter [43, 44] and Hoffmann [17, 18] investigating the fate of nonreplicating antigen using the "network" theory of Jerne [20, 21] to explain tolerance and the normal response. In Richter's model, the humoral response is viewed as a result of a network of interactions of lymphocytes and antibodies resulting in suppression and stimulation of different groups. Hoffmann expanded Richter's model to give a role to the T-cell. These "suppressor" T-cells act much the same as the B-cells in the resulting network except in place of antibody, they produce a monovalent "blocking" substance. Jerne [21] has compared the first model of Bell to those of Richter and Hoffmann.

In both of the network models, the emphasis is on explaining tolerance and no secondary response phenomena is possible. Mohler, Barton and Hsü [35] have also proposed a model explicitly incorporating T-cell - B-cell interactions.

Waltman and Butz [47] have proposed a model of B-cell response to nonreplicating antigen where the threshold nature of the response was used to introduce delays into the differential equations. The numerical testing of this model displays a strong secondary response. Further study of a singular problem which arose in this model was done by Gatica and Waltman [13]. Freedman and Gatica [12] considered a similar model allowing a replicating antigen and studied the stability of critical points and the existence of small periodic solutions.

Bruni, Giovenco, Koch and Strom [4-6] have introduced a model which assumes the B-cells responding to an antigen do so with free energy of binding satisfying a normal distribution. In this way, for low antigen concentrations, only those B-cells with high affinity for the antigen have a good probability of being stimulated. This model has also been compared to the primary response when DNP-BGG was injected into rabbits.

Using a surface to generate the thresholds found in the response, Merrill [31-33] has constructed a model of the humoral response based on a model of the stimulation of B-cells. Induction of both high- and low-zone tolerance was examined in [31]. The stimulation of B-cells by replicating antigen and the periodic solutions of this model were studied in [33] (see also [34] in this volume).

Models of a similar nature which are indications of other types of investigation possible include:

1. Perelson, Mirmirani and Oster [38] who have studied the production of antibody as an optimal strategy problem for a number of simple models of antibody production to gain insight into the possible mechanisms involved. (See also [37].)
2. Lefever and Garay [29] who modeled the immune surveillance against cancerous cells in an attempt to determine the relative importance of this natural defense.
3. Winkelhake [48] used, as a basis, clonal selection and the dichotomy of central and peripheral lymphoid tissue to model some of the processes involved in the specialization of lymphatic cells from pleuri-potent stem cell to antibody-producing cell and despecialization to the memory cell.

The reader is referred to [3] for an indication of the extensive use of mathematical models in theoretical immunology.

A DEEPER LOOK AT SOME OF THE MODELS

The Clonal Selection Model of Bell

A certain small number of B-cells called "target cells" are able to recognize a particular antigen, those that do are stimulated into proliferating when a sufficient amount of antigen is bound to surface receptors. If a very large concentration of antigen is present causing a large percentage of bound receptor sites, the cell will be killed or "tolerized" (made unable to respond) producing high-zone tolerance. The number of bindings of antigen with antibody is assumed to follow the law of mass action, i.e., the number of bindings is proportional to (amount of free antibody) × (amount of free antigen). Let

$N_1(t)$ = number of target (B-) cells per unit volume at time t

$N_2(t)$ = number of proliferating cells (stimulated target cells) per unit volume at time t

$N_3(t)$ = number of plasma cells per unit volume at time t

$N_4(t)$ = number of memory cells per unit volume at time t

$N_5(t)$ = concentration of bivalent (I_gG) antibody, both free and those bound to antigen

$N_6(t)$ = concentration of univalent antigen including free antigen and that bound to antibody and to receptor sites.

Let r' be the fraction of all target cell receptors which are occupied by antigen, and let F(r') be the probability that a target cell is stimulated when r' of the receptors are occupied. In this model, $F(r') = [mr']/[1 + mr']$ where m is the total number of receptors on a target cell. However, some of these cells die, and that happens most frequently if $r' \approx 1$. One choice is simply to let (1 - r') be the fraction of target cells which when stimulated, proliferate rather than die.

The population of target cells is governed by

$$\frac{dN_1}{dt} = -F(r')\frac{N_1}{T_1} + S_1(t) \tag{1}$$

where T_1 = mean time for an optimally stimulated target cell to become a proliferating cell and $S_1(t)$ = source of target cells (from bursal equivalent).

The proliferating cells receive stimulated target cells and lose cells which have begun differentiation, thus the population of proliferating cells is governed by

$$\frac{dN_2}{dt} = (1 - r')F(r')\frac{N_1}{T_1} + H(r')\frac{N_2}{T_2} - \frac{N_2}{T_2'} \tag{2}$$

where $H(r') = [mr' - 1]/[mr' + 1]$ is the net gain (or loss) of proliferating cells due to further proliferation and differentiation into plasma cells and memory cells. (If $r' << 1$, then $H(r') \approx -1$ so that most divisions lead to plasma and memory cells and, if $r' \approx 1$, $H(r') \approx 1$ and most divisions lead to more proliferating cells.) T_2 and T_2' are the mean times for a proliferating cell to divide and die, respectively.

Assuming that a clone produces equal numbers of plasma cells and memory cells, those populations are governed by

$$\frac{dN_3}{dt} = \frac{1 - H(r')}{2} \frac{N_2}{T_2} - \frac{N_3}{T_3} \tag{3}$$

and

$$\frac{dN_4}{dt} = \frac{1 - H(r')}{2} \frac{N_2}{T_2} - \frac{N_4}{T_4} \tag{4}$$

where the quantities $-N_3/T_3$ and $-N_4/T_4$ represent natural death of plasma and memory cells with mean times T_3 and T_4 respectively. If the memory cells are considered functionally equivalent to target cells, equation (1) may be modified by adding a source of new immunocompetent memory cells

$$\frac{1 - H(r')}{2} \frac{N_2}{T_2}$$

The differential equation describing the growth of antibody will have three source terms, the production rate by proliferating cells, $C_2 \cdot N_2$, the production rate by plasma cells, $C_3 \cdot N_3$, and an external source, if any, through passive immunization, $S_2(t)$.

$$\frac{dN_5}{dt} = C_2N_2 + C_3N_3 + S_2(t) - \left[\frac{(2r - r^2)}{T_5} + \frac{1}{T_5'} \right] N_5 \tag{5}$$

where r is the fraction of antibody sites occupied and $\{[2r - r^2]/T_5\} \cdot N_5$ describes the loss of antibody due to binding with antigen and N_5/T_5' is the loss due to natural breakdown at rate $1/T_5'$.

The rate of change of antigen concentration N_6 will be due to the source $S_3(t)$ and natural decay with mean lifetime T_6 and elimination of antigen due to interaction with antibody, elimination occurring in mean time T_5 (antigen-antibody complex eliminated together). Thus,

$$\frac{dN_6}{dt} = S_3(t) - \frac{N_6}{T_6} - \frac{N_6 - L - r'm(N_1 + N_2 + N_4)}{T_5} \tag{6}$$

where $N_6 - L - r'm(N_1 + N_2 + N_4)$ is the concentration of antigen bound to antibody and L is the concentration of free (unbound) antigen and is the unique positive root of the equation

$$N_6 = L + \begin{array}{c}\text{concentration of}\\ \text{antibody sites}\\ \text{bound to antigen}\end{array} + \begin{array}{c}\text{concentration of}\\ \text{receptor sites}\\ \text{bound to antigen}\end{array}$$

$$= L\left(1 + \frac{2kN_5}{1 + kL} + \frac{k'm[N_1 + N_2 + N_4]}{1 + k'L}\right) \tag{7}$$

where k is the intrinsic association constant for antigen-antibody interaction and k' is the constant for antigen-receptor interaction. We can then write

$$r = \frac{kL}{1 + kL}$$

and (8)

$$r' = \frac{k'L}{1 + k'L}$$

The equations (1)-(8) comprise the model for homogeneous antibody production. The model for heterogeneous production of J different antibodies is easily made from this by having equations $(1)_i$-$(5)_i$ model the growth of clone i with intrinsic association constants k_i and k_i'. Equation (6) is then modified to allow for the effects of all the different clones.

$$\frac{dN_6}{dt} = S_3(t) - \frac{N_6}{T_6} - \left(N_6 - L - m \sum_{j=1}^{J} r_j'(N_{1,j} + N_{2,j} + N_{4,j})\right)/T_5 \tag{6'}$$

Up to J = 41 different clones have been handled although computer time becomes prohibitive in simulations. The heterogeneous model shows that towards the end of the primary response, those clones still producing antibody will be those with the highest affinity for the antigen.

The secondary response is displayed due to the larger population of target cells (target cells + memory cells) able to respond to any reintroduction of the antigen after antibody levels are again low. A possible mechanism of low-zone tolerance is suggested by noting that if less than 1/2 of proliferating cells become memory cells, and the stimulation is large enough for only one division, the actual number of target cells able to respond to the antigen will decrease with each introduction of antigen. After long exposure, those B-cells able to respond will be eliminated, resulting in low-zone tolerance.

Predator-prey System as a Model for Growing Antigen

The basis for using a predator-prey framework is that a replicating antigen can be thought of as "food" for the antibody. Of course, destruction of the antigen results in the destruction of the antibody, but the whole process stimulates further antibody production, not unlike the increase in population of predators when prey is available even if the predators do not live to see their progeny. This analogy requires that the antigen under study is "eaten" only by antibody and not greatly affected by the cell-mediated immune system (only one predator), and if the antibody is only stimulated by that antigen (only one prey).

Bell [2] modified the classical Lotka-Volterra predator-prey equations. The Lotka-Volterra equations are

$$\frac{dN_1}{dt} = aN_1 - bN_1N_2$$

$$\frac{dN_2}{dt} = -cN_2 + dN_1N_2$$

where a, b, c, d > 0, N_1 = number of prey, and N_2 = number of predators. The terms N_1N_2 involve the effect of contacts between predators and prey (number of contacts depends upon law of mass action). Bell's model for replicating antigen is

$$\frac{dAg}{dt} = \lambda_1 Ag - \alpha_1 (Ag)_b$$

$$\frac{dAb}{dt} = -\lambda_2 Ab + \alpha_2 (Ab)_b \left(1 - \frac{Ab}{\theta}\right) \tag{9}$$

where Ag is the concentration of antigen, Ab is the concentration of antibody, $(Ag)_b$ is the concentration of bound antigen, and $(Ab)_b$ is the concentration of bound antibody, λ_1, α_1, λ_2, $\alpha_2 > 0$, while θ is the highest possible concentration of antibody in the animal. Assuming one antibody binds with one antigen, i.e., $(Ab)_b = (Ag)_b$, and using the law of mass action, $(Ab)_b = (Ag)_b = k[Ag - (Ag)_b][Ab - (Ab)_b]$.

Solving for $(Ab)_b$ we have

$$(Ab)_b = \frac{kAbAg}{1 + k(Ab + Ag) - k(Ab)_b} \approx \frac{kAbAg}{1 + k(Ab + Ag)}$$

We use the approximation in the model. Substituting this in (9) we have:

$$\frac{dy}{dt} = \lambda_1 y - \alpha_1 \frac{xy}{1 + x + y}$$

$$\frac{dx}{dt} = -\lambda_2 y + \alpha_2 \frac{xy}{1 + x + y}(1 - \lambda x) \tag{10}$$

where $\gamma = 1/k\theta$, $y = Ag$ and $x = Ab$. Make change of variables $s = \int_0^t [1 + x(t') + y(t')]^{-1}dt'$. We obtain (setting $\gamma = 0$ for this discussion)

$$\frac{dy}{ds} = y[\lambda_1 - (\alpha_1 - \lambda_1)x + \lambda_1 y]$$

$$\frac{dx}{ds} = x[-\lambda_2 - \lambda_2 x + (\alpha_2 - \lambda_2)y] \tag{11}$$

Both (10) and (11) make sense biologically only if $\alpha_1 > \lambda_1$ and $\alpha_2 > \lambda_2$ (otherwise y would never decrease and x could never increase).

The results in [2] may be summarized as follows: The phase picture for $x \geq 0$, $y \geq 0$ depends on $R = \alpha_1\alpha_2 - \alpha_1\lambda_2 - \alpha_2\lambda_1$. If $R \leq 0$, Figure 2.1 applies and the antibody (x) response fails to check the proliferation of antigen (y). The origin is a saddle point.

If $R > 0$, there is an additional critical point at the intersection of the lines $y = (1 + x)/(\alpha_2/\lambda_2 - 1)$ and $y = [(\alpha_1/\lambda_1) - 1]x - 1$. The point is $(x_f, y_f) = (\alpha_2\lambda_1/R, \alpha_1\lambda_2/R)$. In this case, if $\alpha_1 > \alpha_2$, (x_f, y_f) is a source, and if $\alpha_2 > \alpha_1$, (x_f, y_f) is a sink. The behavior near (x_f, y_f) depends upon $A = (\alpha_1 - \alpha_2)^2\lambda_1\lambda_2/4\alpha_1\alpha_2$. If $R > A$, (x_f, y_f) is a node.

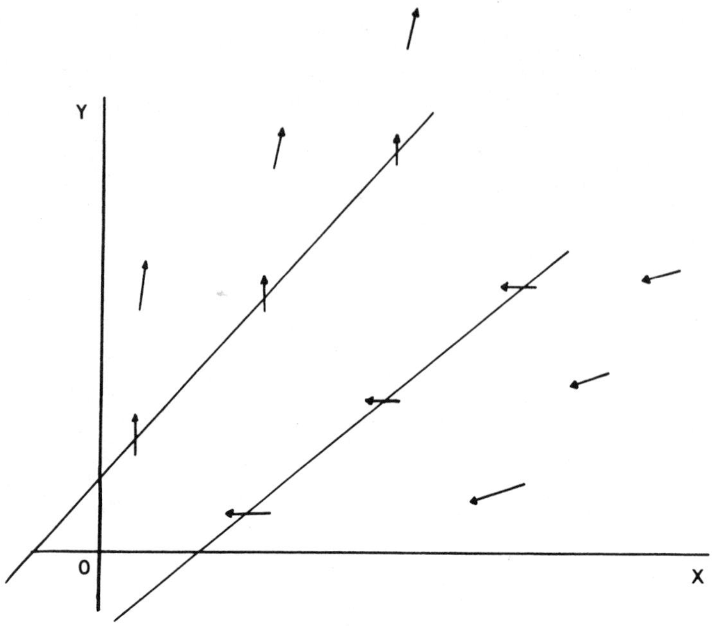

Fig. 2.1. The null curves dx/dt = 0 and dy/dt = 0 if R < 0.

If $\gamma > 0$ (a limit on possible antibody concentration), (11) becomes

$$\frac{dy}{ds} = y[\lambda_1 - (\alpha_1 - \lambda_1)x + \lambda_1 y], \qquad \alpha_1 > \lambda_1$$

$$\frac{dx}{ds} = x[-\lambda_2(1 + x + y) + \alpha_2 y(1 - \gamma x)], \qquad \alpha_2 > \lambda_2 \tag{12}$$

The critical points with $x > 0$, $y > 0$ will be the intersection of $y = (\alpha_1/\lambda_1 - 1)x - 1$ and $y = [\lambda_2(1 + x)]/[\alpha_2(1 - \lambda_2/\alpha_2 - \gamma x)]$. If these curves do not intersect, Figure 2.2 applies and there will be unlimited antigen proliferation as $t \to \infty$, $y \to \infty$, $x \to x_a$.

If the curves intersect twice, as in Figure 2.3, two critical points arise (x_s, y_s), which is a saddle, and (x_f, y_f), a focus or node.

Continuing the study of Bell, G. H. Pimbley [39] investigated the model for periodic solutions bifurcating from the steady-state solution, (x_f, y_f). Consider the system

$$\frac{dx}{ds} = \beta x\left[-\lambda_2 - k(\alpha_2 - \lambda_2)y - \frac{k\alpha_2}{\theta} xy\right]$$

$$\frac{dy}{ds} = y[\lambda_1 - k(\alpha_1 - \lambda_1)x + k\lambda_1 y] \tag{13}$$

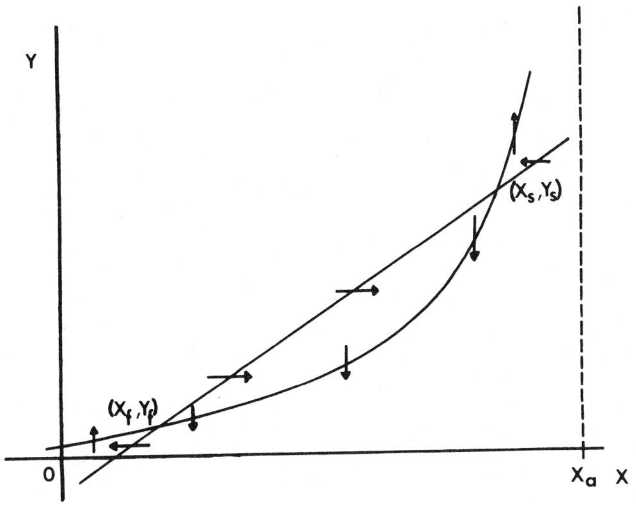

Fig. 2.2.

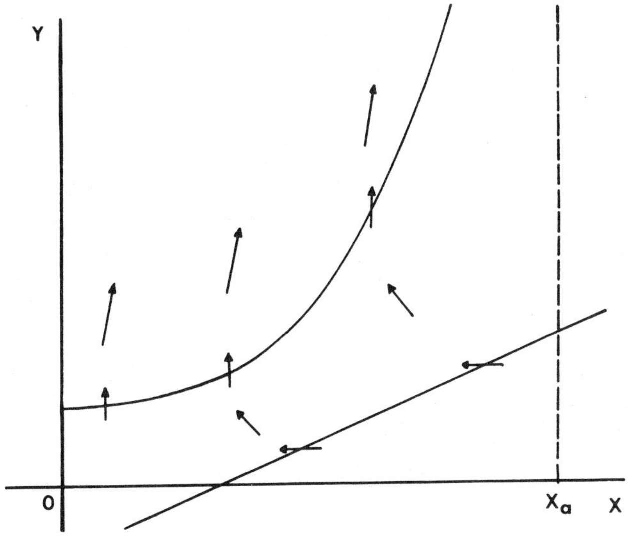

Fig. 2.3.

which is obtained from (10) by making change of variables

$$s = \int_0^t [1 + k(x(t') + y(t'))^{-1}]dt'$$

and adding an auxiliary parameter β.

MAIN THEOREM

1) For system (13) there exists a critical value $\beta_0 = [y_f\lambda_1]/[x_f\lambda_2 + (\alpha_2/\theta)x_f y_f]$. The constant solution (x_f, y_f) is stable if $\beta < \beta_0$. Locally, for $\beta > \beta_0$, the direction field spirals into (x_f, y_f) while for $\beta < \beta_0$ it spirals away from (x_f, y_f).

2) At $\beta = \beta_0$, nontrivial periodic solutions appear, forming a unique continuous one-parameter family in a neighborhood of the constant solution (λ_f, y_f). If $\theta = \infty$ $(\gamma = 0)$ these periodic solutions appear only for $\beta = \beta_0$ as nested dense closed curves in the phase plane, surrounding (x_f, y_f) and extending to ∞ (the saddle point (x_s, y_s) is at ∞ when $\theta = \infty$). If $\theta < \infty$, a unique continuous one-parameter family of periodic solutions branches with small amplitude from the constant solution (x_f, y_f) at β_0, and is defined on some open β-interval, say $\overline{\beta} < \beta < \beta_0$.

3) If $\beta_0 > 1 - (\alpha_2 - \lambda_1 - \lambda_2)(\alpha_2 - \lambda_2)$, the periodic solutions thus generated for $\theta < \infty$ are leftward-branching, supercritical, and asymptotically orbitally stable. There is an exchange of stabilities at β_0 between the constant solution (x_f, y_f) and the supercritical stable continuous branch of periodic solutions. In terms of the natural parameters α_1 and α_2 this bifurcation takes the following form:

4) There exists a value α_{10} of α_1 such that as α_1 is increased through α_{10}, with α_2, λ_1, λ_2 and θ fixed, β_0 increased through unity. Thus, since β-branching of periodic solutions at β_0 is left-branching towards lower β values, α_1-branching of (13) (with $\beta = 1$) is right-branching at α_{10} towards higher α_1 values. We have that $\alpha_{10} > \alpha_2$.

5) Fix a value $\alpha_2^0 > \lambda_1 + \lambda_2$, with λ_1, λ_2 and θ fixed. Let $\alpha_{10} > \alpha_2^0$ be determined as above, and fix a value α_1^0 of α_1 such that $\alpha_1^0 > \alpha_{10}$. Now let α_2 vary, with these fixed parameters α_1^0, λ_1, λ_2, θ. There exists a value α_{20} of α_2 such that, as α_2 is increased through α_{20}, β_0 is decreased through unity. Since β-branching of periodic solutions at β_0 is left-branching, α_2-branching at α_{20} is also left-branching towards lower α_2 values. Thus, α_2 as a parameter under these conditions produces results similar to those produced by β as a parameter.

The proof of this theorem used the Hopf bifurcation theorem [30] and work of Joseph and Sattinger [27]. In Part II. Pimbley investigates the existence of larger amplitude families of periodic solutions. The phase plane is as shown in Figure 2.4.

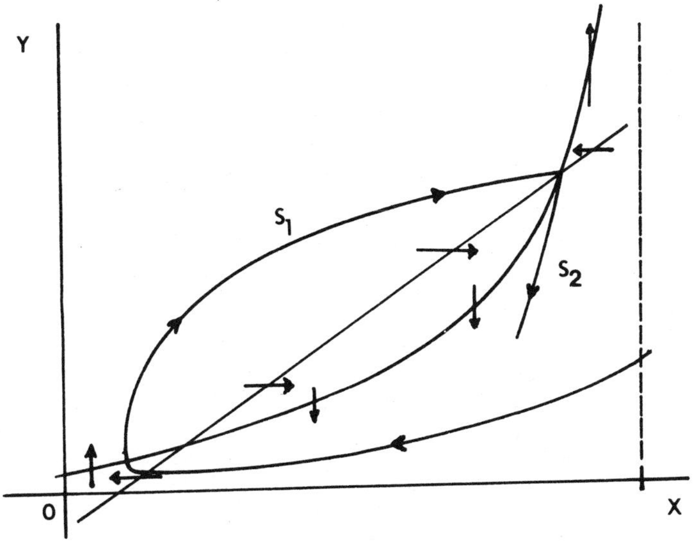

Fig. 2.4.

There exists a value β^* such that at $\beta = \beta^*$, the separatrices S_1 and S_2 in Figure 2.4 unite to form a closed separatrix (if $\theta = \infty$, $\beta^* = \beta_0$). If $\theta < \infty$, set

$$\beta^0 = \frac{\lambda_1 y_s}{\lambda_2 x_s + \frac{\alpha_2}{\theta} x_s y_s}$$

If $\beta^* > \beta^0$, the closed separatrix is stable, i.e., in a sufficiently small inside neighborhood of the loop, all trajectories spiral towards the loop. If $\beta^* < \beta^0$, the loop is unstable. There exists a unique continuous one-parameter family of periodic solutions generated from the closed separatrix at β^*. If $\beta^* > \beta^0$, this family bifurcates from the separation to the right and the periodic solutions are asymptotically orbitally stable. The branching is unstable and to the left if $\beta^* < \beta^0$. Figure 2.5 shows the results of both theorems for $\beta^* > \beta^0$.

This model was modified by Bell in a paper by Pimbley [39]. A third equation governing the concentration of B-cells with a logistic term was added. The result is that the oscillations of increasing amplitude is no longer possible (a very large antibody production would be impossible). This term can be seen an an "environment" term in the predator-prey

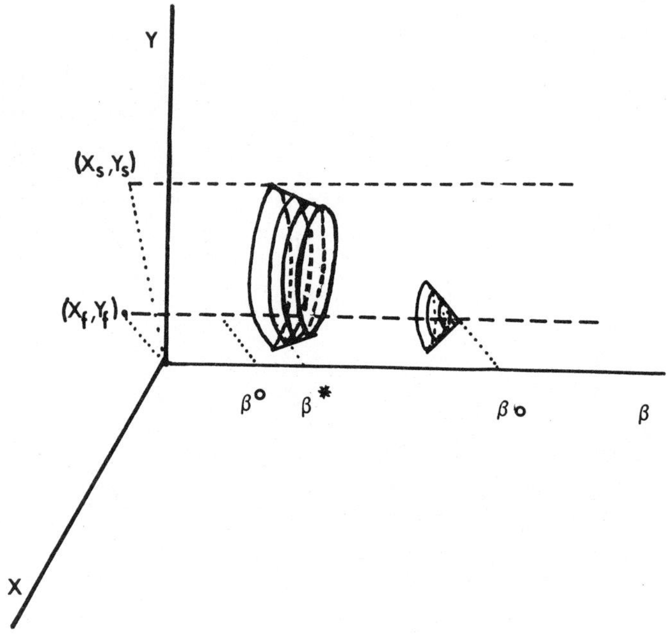

Fig. 2.5.

framework, that is, the outcome of competition between the antibody and antigen also depends on the conditions around them, namely the numbers of B-cells able to respond. The model is

$$\frac{dAg}{dt} = \lambda_1 Ag - \alpha_1 Ab \frac{kAg}{1 + k(Ag + Ab + nC)}$$

$$\frac{dAb}{dt} = -\lambda_2 Ab - \alpha_2 Ab \frac{kAg}{1 + k(Ag + Ab + nC)} + \gamma C \frac{kAg}{1 + k(Ag + Ab + nC)} \quad (14)$$

$$\frac{dC}{dt} = -\lambda_3 C + \alpha_3 C \frac{kAg}{1 + k(Ag + Ab + nC)} \left(1 - \frac{C}{\theta}\right) + S$$

where C is the concentration of B-cells, S is a constant source term, λ_3 the rate of decay of lymphocytes, α_3 maximum proliferation rate of B-cells when stimulated, and θ is the maximum number of B-cells possible, n is the number of receptors on a B-cell, and γ is maximum production rate of antibody.

Making the change of variables

$$s = \int_0^t [1 + k(u(t') + v(t') + nw(t'))]dt'$$

and setting u = Ag, v = Ab, and w = c we have

$$\frac{du}{ds} = u[\lambda_1 + k\lambda_1 u - k(\alpha_1 - \lambda_1)v + kn\lambda_1 w]$$

$$\frac{dv}{ds} = v[-\lambda_2 - k(\alpha_2 + \lambda_2)u - k\lambda_2 v - kn\lambda_2 w] + k\lambda uw \tag{15}$$

$$\frac{dw}{ds} = w\left[-\lambda_3 + k(\alpha_3 - \lambda_3)u - k\lambda_3 v - kn\lambda_3 w - \frac{k\alpha_3}{\theta}uw\right] + S[1 + k(u + v + nw)].$$

A necessary condition that the three null surfaces du/ds = 0, dv/ds = 0 and dw/ds = 0 intersect in two critical points (u_f, v_f, w_f) and (u_s, v_s, w_s) in the first octant is that

$$\frac{\alpha_1 - \lambda_1}{\lambda_1} > \frac{\alpha_3}{\alpha_3 - \lambda_3}, \qquad \frac{\alpha_1 - \lambda_1}{\lambda_1} > \frac{n\lambda_2}{\gamma}$$

with $\alpha_1 > \lambda_1$ and $\alpha_3 > \lambda_3$. The surface dw/ds = 0 is an invariant plane when $\lambda_3 = 0$ and s = 0, so that in this case (15) can be reduced to a second-order system

$$\frac{du}{ds} = u[\lambda_1(1 + kn\theta) + k\lambda_1 u - k(\alpha_1 - \lambda_1)v]$$

$$\tag{16}$$

$$\frac{dv}{ds} = k\theta\gamma u + v[-\lambda_2(1 + kn\theta) - k(\alpha_2 + \lambda_2)u - k\lambda_2 v]$$

in the w = θ plane. The phase picture in this case looks similar to Figure 2.4 in the earlier model.

More generally, when S = 0, inserting an auxiliary parameter β in the second equation of (15) as before, Pimbley proves the existence of a critical value $\beta_c > 0$ so that for $\beta > \beta_c$, (u_f, v_f, w_f) is a stable focus and for $\beta < \beta_c$, (u_f, v_f, w_f) is a saddle focus. At $\beta = \beta_c$, nontrivial periodic solutions branch from the constant solution (u_f, v_f, w_f) forming a unique one-parameter family of periodic solutions. A discussion of the stability of the bifurcation for $\lambda_3 > 0$ is left unfinished due to the loss of the special techniques available in two-dimensions. As in the earlier model, a translation to bifurcation with a natural parameter is made (with α_1 as the bifurcation parameter).

It is noted that for small λ_2, λ_3, γ and α_2 most trajectories in phase space diverge from the steady state solution (u_f, v_f, w_f). Only when the elimination rate of bound antigen, α_1, is large enough do some bounded trajectories tend from the unstable family of periodic solutions to the steady state, thus in this model, an infectious attack gets quickly out of hand unless α_1 is large enough and the initial conditions are near the steady state. Extensions and fine-tuning of these results are summarized in [42].

A paper by Hsü and Kazarinoff [19] gives further bifurcation results of this model, most notably giving explicit criteria for existence and stability of bifurcating solutions.

Network Models: The Models of Richter and Hoffmann

THE MODEL OF RICHTER

The immune response observed is a result of the interaction of lymphocytes and antibodies resulting in suppression and stimulation of different groups of both. If an antigen, Ag, is present, it elicits the production of a heterogeneous class of antibody called Ab_1, the idotypes present on antibodies in Ab_1 are recognized by another heterogeneous collection called Ab_2. If the levels of Ab_1 rise enough to stimulate B-cells with Ab_2 receptors, Ab_2 will be produced and released. Presence of Ab_2 will inhibit Ab_1 combination with Ag. Similarly, if Ab_2 levels become high enough, Ab_3 production is stimulated. Presence of Ab_3 will inhibit Ab_2-Ab_1 combinations, allowing Ab_1 to again combine with Ag. The following shows the possible results of this network regulation:

1. low-level response: $Ag \xrightarrow{\text{stimulates}} Ab_1$
   Ab_1 levels below threshold for Ab_2 stimulation,

2. low-zone tolerance: $Ag \xrightarrow{stim} Ab_1 \xrightarrow{stim} Ab_2$
   Ab_2 inhibits Ab_1-Ag combination,
3. normal response: $Ag \xrightarrow{stim} Ab_1 \xrightarrow{stim} Ab_2 \xrightarrow{stim} Ab_3$
   Ab_3 eliminates Ab_2 inhibition of Ab_1-Ag combination,
4. high-zone tolerance: $Ag \xrightarrow{stim} Ab_1 \xrightarrow{stim} Ab_2 \xrightarrow{stim} Ab_3 \xrightarrow{stim} Ab_4$
   Ab_4 eliminates Ab_3, response appears as in low-zone tolerance.

Richter's model has a simple form. Assume that the ratio of B-lymphocytes with a given specificity to antibody of that same specificity is constant, thus one equation will suffice to govern numbers of B-cells and concentration of antibody of each type.

Let X_i be the B-cell population with receptors of type Ab_i with respect to some antigen (concentration of antigen is X_0), then

$$\frac{dX_i}{dt} = \left\{\frac{1}{\tau_b} f_b(X_{i-1}, X_{i+1}) - \frac{1}{\tau_d} f_d(X_{i-1}, X_{i+1})\right\} X_i$$

where f_b and f_d are fractions of B-cells of type X_i stimulated due to presence of X_{i-1} and killed due to presence of X_{i+1}, respectively. Both are functions of X_{i-1} and X_{i+1} as X_{i+1} inhibits stimulation due to X_{i-1}. τ_b and τ_d are the mean times for cell replication and cell destruction, respectively due to X_{i-1} and X_{i+1}.

Assume all bindings occur with identical binding constant K, then if we assume that no antibody can simultaneously use its binding site and be bound at its idiotope, letting $S_i = K \cdot \alpha X_i = K \cdot$(free antibody concentration), we can write

$$f_b(S_{i-1}, S_{i+1}) = 1 - \frac{\left[\sum_{k=0}^{\mu} \binom{M}{k} \left(\frac{S_{i-1}}{1 + S_{i+1}}\right)\right]^k}{\left(1 + \frac{S_{i-1}}{1 + S_{i+1}}\right)^M}$$

where a lymphocyte is stimulated if μ of its M receptors are occupied.

Letting $y_i = S_{i-1}/1 + S_{i+1}$, we can write

$$f_b(y_i) = 1 - \frac{\left[\sum_{k=0}^{\mu} \binom{M}{k} y_i^k\right]}{(1 + y_i)^M} = \left[\sum_{k=1}^{M} \binom{M}{k} y_i^k\right] / (1 + y_i)^M$$

Similarly, setting $z_i = (S_{i+1})/(1 + S_{i-1})$,

$$f_d(z_i) = \left(\sum_{k=1}^{M} \binom{M}{k} z_i\right) / (i + z_i)^M$$

where a B-cell is killed if μ of its M receptors are occupied. (Note: The antigen equation has $f_b = 0$ as there is nothing of species -1.)

THE MODEL OF HOFFMANN

Taking the simple network model of Richter, Hoffmann has put it in the framework of present knowledge, expanding the model to contain these effects. Hoffmann's model is only one of two which gives a role in the response to the T-cell. These "suppressor" T-cells act much the same as the B-cells in Richter's model except that they are more easily stimulated and in place of antibody, produce a monovalent "blocking" substance. This model assumes that the antigen binding site is the same as the idiotype, thus any suppression is a result of competition for the binding sites.

Hoffmann states the network in terms of "positive" and "negative" T- and B-cell populations. The antigen (Ag_-) stimulates a population of T-cells (T_+) and B-cells (B_+). These stimulate the populations of suppressor cells T_- and B_-. T_- produces a monovalent substance that blocks + receptors and sticks to phagocytic cells causing specific killing of anything with + receptors. B_- produces antibody which binds to + receptors on lymphocytes and free antibody.

In the *virgin state*, all quantities T_+, T_-, B_+, B_- are low and the network is non-operational. The *suppressed state* results when there are high T-cell populations T_+ and T_- whose monovalent substances block the receptors on both populations of B-cells. This can result because the T-cell is more easily stimulated, and at low antigen concentrations, only T-cell clones would grow, resulting in low-zone tolerance. The *immune state* is demonstrated when the + populations are not suppressed. This results when a high concentration of antigen causes rapid growth in T_+, whose monovalent substance is able to block the growth of T- and B- populations.

The form of the equations in Hoffmann's model is very simple; all are of the form

$$\frac{dx}{dt} = X(R - D_k - D_{Ab} - D_{ns}) + I_x$$

where X is the quantity T_+, T_-, B_+, B_-, the term I_X is the constant "immigration" term, i.e., source of new lymphocytes from bone marrow or thymus.

R is the replication rate, D_K and D_{Ab} are rates of specific killing by cytotoxic effector cells and by antibodies and D_{ns} is the constant non-specific death rate. All the quantities R, D_K, D_{Ab} and D_{ns} are functions of the concentrations of other sign, that is, when X is T_+, R is a function of numbers of T_- and B_-. The rates of B- and T-cell replications are assumed to be different while rate of B-cell killing is the same as T-cell rate of same sign (probably not a good assumption, as C3 receptors have been found on B-cells and not T-cells so that B-cells probably lyse faster). The functions R, D_K, D_{Ab} will be defined in terms of a threshold function $\phi(x,y) = 1/[1 + (y/x)]$. For instance, R for T_+ as stimulated by T_- and B_- is $(1/\tau) \cdot \phi$ (concentration of "-" receptors, amt. of blocking).

The antigen is governed by $dAg/dt = 1/\tau_{Ag} \cdot Ag$ where τ_{Ag} is constant rate of decay. In this model, note that the immune system is assumed to not have any effect on the growth of the antigen, thus the model demonstrates only the dynamics of the network.

The model has undergone minor changes and extensions as noted in [18].

A Threshold Model

THE MODEL OF WALTMAN AND BUTZ

A small challenge of antigen will produce little or no immune response, the antigen is either destroyed by nonspecific agents or by natural decay or by combination with antibody already produced. When the concentration of antigen reaches a level high enough, stimulation of B-cells occurs.

Let

$x(t)$ = concentration of free (unbound) antigen

$y(t)$ = concentration of free (unbound) receptors on B-cells able to recognize the antigen

$z(t)$ = concentration of free antibody at time t (not on surface of B-cell)

$w(t)$ = concentration of antigen bound to receptors

and let θ_1 be the threshold prerequisite for proliferation (probably in terms of number of receptors cross-linked in capping phenomena). Let $f_1(x,y,z)$ be a function which describes the stimulation of a B-cell so that when the "integrated" stimulation

$$F(t) = \int_{\tau_1(t)}^{t} f_1(x(s),\ y(s),\ w(s))ds \quad \text{is} \quad \theta_1,$$

the B-cell will start proliferation at time t. Here, $\tau_1(t)$ is the time stimulation by antigen was started, resulting in proliferation at time t. The stimulation of a lymphocyte at time t does not happen instantaneously, the stimulation depends on antigen and lymphocyte concentrations at previous times, starting at $\tau_1(t)$.

Let $L(\tau)$ be the B-cell concentration at time τ and the duration of the proliferation phase given by

$$\int_{\tau_2(t)}^{t} f_2(L(u))du = \theta_2,$$

i.e., proliferation which started at $\tau_2(t)$ will terminate at time t with plasma cells ready to produce antibody if the clone size is large enough. $f_2(L)$ is a function which describes this process. (In numerical tests of this model, $f_1(x,y,w) = x \cdot y + w$ and $f_2(L) = L$.)

Assuming $L(\tau)$ is proportional to $y(t) + w(t)$ we have

antigen: $dx/dt = -rx(t)y(t) - sx(t)z(t)$

The free antigen is bound by receptors and free antibody following the law of mass action.

receptors: $dy/dt = -rx(t)y(t) + \alpha rx(\tau_1(t))y(\tau_1(t))H(t - t_1)$

Free receptor concentration is decreased by combination with antigen and increased proportional to the stimulation in the past.

antibody: $dz/dt = -sx(t)z(t) - \gamma x(t) + \beta rx(\tau_2(t))y(\tau_2(t))H(t - t_2)$

Antibody is reduced by binding with antigen and exponential decay, and increased proportional to the number of plasma cells mature at time t (those starting proliferation after $\tau_2(t)$ will not be mature at time t).

antigen bound to receptors: $dw/dt = rx(t)y(t)$

$$H(t) = \begin{cases} 1 & \text{if} \quad t \geq 0 \\ 0 & \text{if} \quad t < 0 \end{cases}$$

The decays are defined by:

$\tau_1(t)$ is such that $\int_{\tau_1(t)}^{t} f_1(x(s), y(s), w(s))ds = \theta_1, \quad t \geq t_1$

$\tau_1(t) = 0$ if $t \leq t_1$

$\tau_2(t)$ is such that $\int_{\tau_2(t)}^{t} f_2(y(s) + w(s))ds = \theta_2, \quad t \geq t_2$

$\tau_2(t) = 0$ if $t \leq t_2$

where t_1 and t_2 are given by

$$\int_0^{t_1} f_1(x,y,w)ds = \theta_1 \quad \text{and} \quad \int_0^{t_2} f_2(y + w)ds = \theta_2$$

or by $t_1 = +\infty$, $t_2 = +\infty$ if no such t_i's exist. (t_1 is the first possible time of lymphocyte proliferation and t_2 is initial instant of free antibody secretion by plasma cells.)

A small number of numerical tests were run on the model for $f_1(x,y,w) = xy + w$ (triggering depends on the amount of antigen bound to receptors, $w(t)$, and on the rate of this binding, x, $y = w'/r$) and $f_2(\eta) = \eta$. The main feature of the numerical tests is the strong, quick secondary response. This is a result of a larger pool of B-cells able to recognize the antigen. (In this model, the quantity $y(t)$ is assumed proportional to all virgin B-cells, proliferating cells, plasma cells, and memory cells of a given specificity so the memory cell population is demonstrated by a large y value after the primary response.)

It is shown that the model itself is well-posed as long as f_1 is continuous and locally Lipschitzian, $f_1(x,y,w) > 0$ if $x > 0$ and $y > 0$, f_2 continuous, locally Lipschitzian and $f_2(p) > 0$ if $p > 0$. So more general f_1 and f_2 are possible.

A singular problem of mathematical interest results when the biologically attractive assumption $f_1 = w$ = concentration of antigen bound to receptors is used. It is a singularity in the sense that, since $w(0) = 0$, the function multiplying the derivative is zero at the initial point.

In [13], the existence and uniqueness of the solution was established for this case.

THE RESPONSE TO REPLICATING ANTIGEN: FREEDMAN AND GATICA

Thresholds of the type used by Waltman and Butz were combined with more general interaction and growth functions and a replicating antigen by Freedman and Gatica [12].

Antigen capable of replicating is injected at time zero to the threshold time $\hat{t}$. Antigen concentration grows and binds (irreversibly) to B-cell receptors.

x, y, z, w are defined as Waltman and Butz except that $w(t)$ is the concentration of antigen bound either to receptors or free antibody.

Then for $0 \leq t \leq \hat{t}$,

$$\frac{dx}{dt} = x(t)g(x(t)) - \beta x(t)y(t)$$

$$\frac{dy}{dt} = -\beta x(t)y(t)$$

$$\frac{dz}{dt} = 0$$

$$\frac{dw}{dt} = \beta x(t)y(t) - \varepsilon w(t)$$

(includes possible loss of bound antigen at rate ε)

$$x(0) = x_0 > 0, \ y(0) = y_0 > 0, \ z(0) = w(0) = 0$$

where $g(x)$, a function with

$$g(0) = \alpha > 0, \ \frac{dg}{dx}(x) \leq 0 \quad \text{for} \quad x \geq 0$$

is the antigen's specific growth rate with only density dependent hinderance.

The threshold time $\hat{t}$ is defined in a way analogous to Waltman and Butz

$$\int_0^{\hat{t}} \hat{f}(x(t),\ y(t),\ w(t)dt = \hat{m} \quad \text{for some function } \hat{f}$$

and number $\hat{m}$.

If a time $\hat{t}$ exists, receptors are produced. Let

$$\hat{H}(t) = \begin{cases} 0 & \text{for } t < \hat{t} \\ 1 & \text{for } t \geq \hat{t} \end{cases}$$

then the second equation can be written

$$\frac{dy}{dt} = -\beta x(t)y(t) + y(t)h(y(t))\hat{H}(t)$$

where the growth function h has the same properties as g above.

If there is a time t* satisfying

$$\int_0^{t^*} f^*(x(t),\ y(t),\ z(t),\ w(t))dt = m^*$$

for some function $f^* \geq 0$ and number m*, free antibody will be produced. Antigen challenges resulting in no t* would then be a subthreshold for a primary response.

Define

$$H^* = \begin{cases} 0 & \text{for } t < t^* \\ 1 & \text{for } t \geq t^* \end{cases}$$

then if a primary response is to result (a t* exists), free antibody is produced by growth rate h(y(t)), catabolized by rate δ and combines with antigen by mass action at rate γ. The model becomes

$$\frac{dx}{dt} = xg(x) - \beta xy - \gamma xz$$

$$\frac{dy}{dt} = -\beta xy + yh(y)\hat{H}(t)$$

$$\frac{dz}{dt} = -\gamma xz - \delta z + k(y)H^*(t)$$

$$\frac{dw}{dt} = \gamma xy + \delta xz - w$$

where $k(0) = 0$, $(dk)/(dy)(y) \geq 0$ for $y \geq 0$ and all constants are positive.

Considering the dynamics of the above system of differential equations for $t > t^* > t$, $H(t) = H^*(t) = 1$, the resulting model was examined for several natural choices for the growth functions g, h, k. In at least one example, Hopf bifurcation results in small periodic solutions.

This model contains the thresholds but not the delays (and associated complications) of the Waltman-Butz model. The possibility of a periodic solution along which the thresholds switch on and off is a reasonable conjecture.

The Model of Bruni, Giovenco, Koch, and Strom

It is assumed that there is a continuous distribution of immunocompetent B-cells with respect to the association constant for a given antigen and the distribution is Gaussian with mean 0 and variance σ in the free energy of binding between the antigen and cell receptors. The distribution is limited to a finite interval of K values, K_1 to K_2. Stimulation of B-cells into plasma and memory cells occurs when the number of receptors occupied per cell by the antigen falls between n_1 and n_2. Less than n_1 bindings result in no response as in the threshold model, while more than n_2 bindings results in reversible "tolerization."

Let $R(K,t)$ and $L(K,t)$ be the number of free and bound receptors on cell surfaces respectively. For an antigen concentration $H(t)$ and for each value of the association constant $K = L(K,t)/[R(K,t) \cdot H(t)]$ between K_1 and K_2, the probability of a receptor of affinity K is occupied is

$$p(K,H) = \frac{L(K,t)}{R(K,t) + L(K,t)} = \frac{K \cdot H(t)}{1 + KH(t)}$$

so that the probability that a cell with affinity K is stimulated is

$$\left(\sum_{n=n_1}^{n_2} \binom{m}{n} (K \cdot H)^n\right) / (1 + KH)^m$$

as in Richter's model (where $K \cdot H = y_i$). The model consists of four differential equations for each fixed value of K, and one equation governing the antigen concentrations.

immunocompetent B-cells:

$$\frac{\partial C(K,t)}{\partial t} = \alpha_c \frac{1 - KH}{1 + KH} p_s(KH)C(K,t) - \frac{1}{\tau_c} C(K,t) + p_c(K)$$

= (increase due to new memory cells - loss of virgin cells in proliferation) - natural death + new lymphocytes from stem cells.

plasma cells:

$$\frac{\partial C_p(K,t)}{\partial t} = 2\alpha_c \frac{KH}{1 + KH} p_s(KH)C(K,t) - \frac{1}{\tau_p} C_p(K,t)$$

= plasma cells from proliferation - natural death.

Assuming that memory cells also produce free antibody for a time after their generation, let I(K,t) be the "intensity" of antibody production by memory cells of association constant K (I(K,t) is given an integral form in the model), then

antibody sites:
(2 per molecule)

$$\frac{\partial S(K,t)}{\partial t} = \alpha_s C_p(K,t) + I(K,t) + \alpha_s' C(K,t) - k(K)S(K,t)H(t) + k'(K)B(K,t) - \frac{1}{\tau_s} S(K,t)$$

= production by plasma cells + production by memory cells + basal production by other cells - sites lost by binding to antigen + sites gained by dissociation of the complexes - natural breakdown

where k and k' are association and dissociation rates of the reaction:

$$\text{antibody} + \text{antigen} \underset{k'}{\overset{k}{\rightleftharpoons}} \text{antibody - antigen complexes}$$

complexes:

$$\frac{\partial B(K,t)}{\partial t} = k(K)S(K,t)h(t) - k'(K)B(K,t) - \frac{1}{\tau_B} B(K,t)$$

and

antigen concentration:

$$\frac{dH}{dt} = -H(t)\int_{K_1}^{K_2} k(K)S(K,t)dK + \int_{K_1}^{K_2} k'(K)B(K,t)dK - \frac{1}{\tau_H} H(t) + A(t)$$

where A(t) is the antigen put in the system at time t.

The model coefficients were set to approximate the response to DNP-BGG in rabbits, the best results being in measuring the change in the average association constant as the challenge is discovered and then defeated.

Further mathematical examination of the model verifies existence and uniqueness of solutions and reveals stability properties in a simplifying case.

Useful results on bilinear systems [5] make it possible to establish the global asymptotic stability of the unique equilibrium for a special case.

T-cell: B-cell Model of Mohler, Barton and Hsü

Except for the case of T-independent antigens, the T-cell plays a crucial role in the humoral response. In modeling the T-cell influence on the B-cell, Mohler et al. [35] have assumed that the T-cell reacts to antigen in a way similar to the B-cell, except that instead of producing antibody, an antibody-like molecule (see [14]) "IgT" is produced which aids (or inhibits) the stimulation of a B-cell. IgT-Ag complexes bind via the common (F_c) region to macrophages. These then bind through antigenic determinants to the B-cell. (The "processing" involves enabling the macrophage to bind antigen and present it to the B-cell.) IgT-Ag complexes which do not have a macrophage present cause *inhibition* of this stimulation sequence (due to a missing required macrophage cell-product?).

Let

x_1 = population density of B-cells able to recognize a partigular Ag
x_2 = density of plasma cells
x_3 = density of unbound antibody sites
x_4 = density of immune complexes
$h(t)$ = unbound antigen concentration

The construction of the model equation depends on the definition of two stochastic quantities:

P_d = probability that a B-cell differentiates to a plasma cell (dependent on $h(t)$ and the association constant k)

and

P_s = probability that antigen stimulates the cell (also h and k dependent).

As in several of the other models, these quantities are approximated by the familiar functions

$$P_d = \frac{kh}{1 + kh}$$

and

$$P_s = \begin{cases} 1 & \text{for } \gamma_1 \leq kh \leq \gamma_2 \\ 0 & \text{for other } kh \end{cases}$$

Now letting

$$u_1 = P_s(1 - 2P_d) = \frac{dx_1}{dt} = \alpha u_1 x_1 - \frac{x_1}{\tau_1} + \beta_k$$

= (proliferation of B-cells - loss of plasma cells) - death of B-cells and source (bone marrow)

$$\frac{dx_2}{dt} = 2\alpha u_2 x_1 - \frac{x_2}{\tau_2} \quad \text{where} \quad u_2 = P_s P_d$$

= gain of plasma cells through differentiation - death

$$\frac{dx_3}{dt} = -C_k K h x_3 - \frac{x_3}{\Sigma_3} + \alpha' x_2 + C_k x_4 + \alpha'' x_1$$

+ loss of free sites (depending on k) by combining with Ag − natural loss − production from plasma cells

+ dissociation of Ab-Ag complexes + production from other B-cells

$$\frac{dx_4}{dt} = C_k k h x_3 - D_k + \frac{1}{t_4} x_4$$

= Ag-Ab complexes - (loss by disassociation and elimination).

These equations, along with the equation for h,

$$\frac{dh}{dt} = \dot{h}_i - \frac{h}{\tau_h} - \sum_h \frac{x_4}{\tau_4}$$

form a model for the response with input of a T-independent antigen $h_i(t)$.

To allow for T-cell influences in all other cases, a similar set of four equations (x_{1T} - x_{4T}) governing the density of T-cells T-"plasma" cells (producing IgT), IgT and IgT-Ag complexes is added to the model. It only remains to quantize the effect of IgT on the B-cells. To simplify the model, Moher et al. assume there is no shortage of macrophages, but they have limited numbers of (F_c) receptors.

The first B-cell equation is the same, but the coefficients P_s and P_d become:

$$P_s = \begin{cases} 1 & \gamma_1' \leq h'h + \gamma_T \dfrac{x_{4T}}{x_3} \leq \gamma_2', \; x_3 > 0 \\ 0 & \text{elsewhere} \end{cases}$$

and

$$P_d = \begin{cases} \dfrac{k'h + x_{4T}/x_3}{1 + k'h + \gamma_T(x_{4T}/x_3)} & x_{4T} \leq x_{4m} \quad (x_3 > 0) \\[2ex] \dfrac{k'h + \gamma_T(2x_{4m} - x_{4T})/x_3}{1 + k'h + \gamma_T(x_{4T}/x_3)} & x_{4T} \geq x_{4m} \quad (x_3 > 0) \end{cases}$$

where k' is the affinity of IgT for antigen, γ_1', γ_2' are the threshold values for stimulation.

γ_T is a T-B "coupling coefficient" to allow for increased stimulation of B-cell activity.

x_{4m} represents the finite limitation of bindings with macrophage surface F_c receptors.

A laboratory research program has been started to verify and/or provide data for improvement of the model.

The Stimulation Model of Merrill

The underlying assumption here is that there is a regulatory mechanism within the B-cell which responds to antigen on surface receptors in sufficient concentration. It is also assumed that the size of the clone depends on sensing the concentration of antigen on surface receptors, that is, antigen is required not only for the initial stimulation but through the replicating (clonal) phase and the differentiation phase.

Following from these assumptions and the threshold features of the response, the threshold is included in the model by requiring solutions to stay on or near a particular surface in this model given by
$x^3 + [a - (1/2)]x + b - 1/2 = 0$ where

$$x = \frac{\text{\# of B-cells unstimulated} - \text{\# of B-cells stimulated}}{\text{total \# of B-cells}}$$

is a measure of stimulation, and

a = concentration of free antibody

b = concentration of free antigen

This surface is called the cusp catastrophe [45] and generates both a switching "on" and a switching "off" at different threshold values. It generates a threshold curve shown in Figure 2.6. To the right of the curve B-cells are rapidly stimulated, to the left the stimulation is marginal (only cells with large affinity for the antigen).

Using standard mass action interactions for antibody and antigen, the model becomes

$$\varepsilon \frac{dx}{dt} = -\left(x^3 + \left(a - \frac{1}{2}\right)x + b - \frac{1}{2}\right) \tag{17}$$

$$\frac{da}{dt} = \begin{array}{c}\text{antibody production}\\ \text{(proportional to}\\ \text{stimulation)}\end{array} - \begin{array}{c}\text{natural}\\ \text{degrading}\\ \text{of antibody}\end{array} - \begin{array}{c}\text{loss of antibody}\\ \text{binding with}\\ \text{antigen}\end{array}$$

$$= \delta \frac{1 - x}{2} - a - \gamma_1 ab$$

$$\frac{db}{dt} = \begin{array}{c}\text{loss of antigen}\\ \text{binding with antibody}\end{array} + \begin{array}{c}\text{growth (or decay)}\\ \text{of antigen}\end{array}$$

$$= -\gamma_1 ab + \gamma_2 b$$

$$-1 \le x(0) \le 1,\ a(0) \ge 0,\ b(0) \ge 0.$$

In the first equation, ε is a small positive number indicating the rapid stimulation action compared to the slower action of the rest of the response (which is scaled to make the second equation simple - 1 time unit = half-life of antibody being considered).

This system is in the form of a singularly perturbed initial value problem and information on the qualitative features of the model. If the antigen is not able to replicate ($\gamma_2 < 0$) in [32], $x = 1$, $a = 0$, $b = 0$ is shown to be a global attractor if δ is not too large. In [33] periodic solutions are shown to exist and one method, setting $\varepsilon = 0$ in (17) and studying the dynamics on the resulting (2-dimensional) surface is reminiscent of Pimbley's work. (See also [34].)

The model was expanded to a model of the humoral response by altering the first equation and adding one governing the population of small B-cells, c(t). That model was

$$\varepsilon \frac{dx}{dt} = -\left(x^3 + \left(a - \frac{1}{2}\right)x + bc - \frac{1}{2}\right)$$

$$\frac{da}{dt} = \delta \frac{1 - x}{2} - a - \gamma_1 ab$$

$$\frac{db}{dt} = -\gamma_1 ab + \gamma_2 b \tag{18}$$

$$\frac{dc}{dt} = \gamma_3 c(c_{max} - c)(a - b)$$

$$-1 \le x(0) \le 1,\ a(0) \ge 0,\ b(0) \ge 0,\ 0 \le c(0) \le c_{max}$$

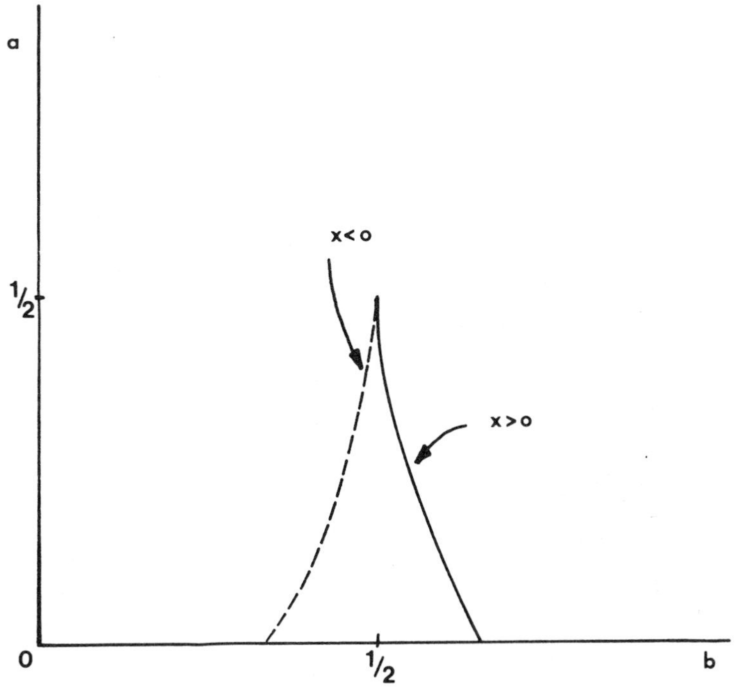

Fig. 2.6. Curves in a-b plane expressing the threshold for the response being activated (the solid line) and inactivated (the dashed line).

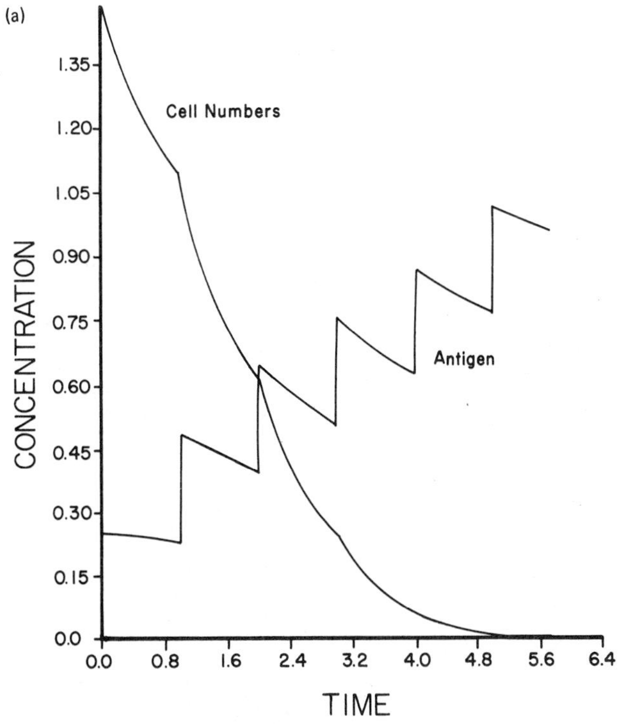

Fig. 2.7a

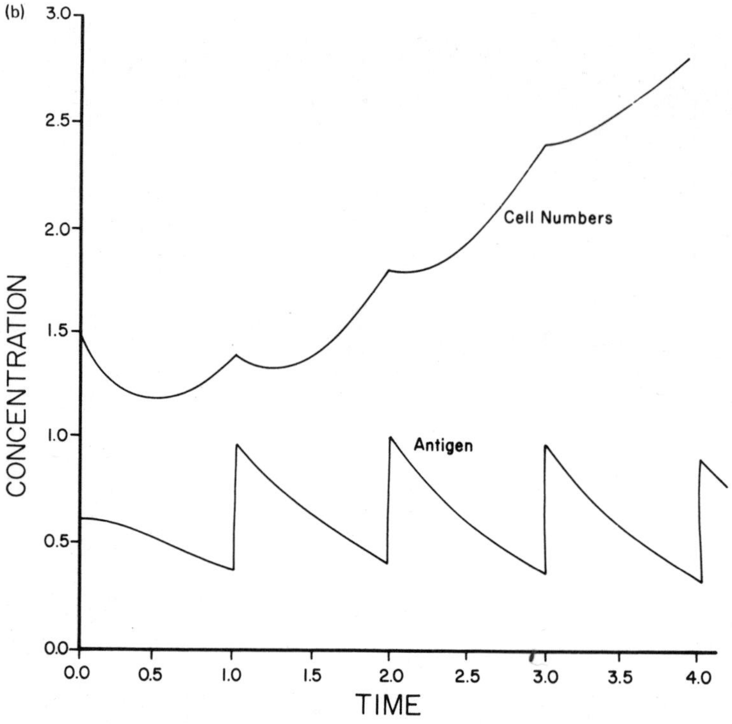

Fig. 2.7b.

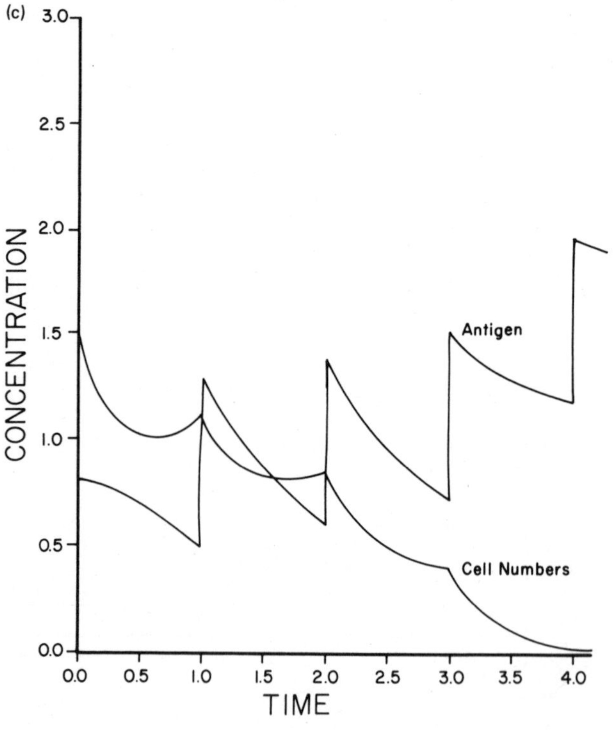

Fig. 2.7c

Numerical simulations [31] of (18) show a primary and secondary response and some of the varied data concerning the induction of tolerance are displayed in Figures 2.7a, b, c.

ACKNOWLEDGMENTS

This is an update of the same title published as Technical Report IM76-1, Department of Mathematics, University of Iowa, Iowa City, Iowa, 1976, which was supported by Public Health Service Grant IRO1CA 18639-01 from the National Cancer Institute.

REFERENCES

1. G. I. Bell, "Mathematical model of clonal selection and antibody production," in three parts, *J. Theor. Biol.* 29(1970), 191-232; 33(1971), 339-378; 33(1971), 379-398.

2. G. I. Bell, "Predator-prey equations simulating an immune response," *Math. Biosci.* 16(1973), 291-314.

3. G. I. Bell, A. S. Perelson and G. H. Pimbley, Jr. (Eds.) *Theoretical Immunology*, Marcel Dekker, New York (1978).

4. C. Bruni, M. A. Giovenco, G. Koch and R. Strom, "A dynamical model of humoral immune response," *Math. Biosci.* 27(1975), 191-211.

5. C. Bruni, M. A. Giovenco, G. Koch and R. Strom, "The immune response as a variable structure system," in *Variable Structure Systems with Application to Economics and Biology*, A. Ruberti and R. R. Mohler (Eds.), Springer-Verlag, Berlin (1975), 244-264.

6. C. Bruni, M. A. Giovenco, G. Koch and R. Strom, "Modeling of the immune response: a system approach," in *Theoretical Immunology*, G. I. Bell, A. S. Perelson and G. H. Pimbley, Jr. (Eds.), Marcel Dekker, New York (1978), 379-414.

7. F. M. Burnet, *The clonal selection theory of immunity*, Vanderbilt U. Press, Nashville, Tenn. (1959).

8. S. Cohen, "A model for the mechanism of antibody induction and tolerance, with specific attention to the affinity characteristics of antibody produced during the immune response," *J. Theor. Biol.* 27(1970), 19-29.

9. S. Cohen, "A quantitative model for mechanisms of antibody formation and tolerance," in Adv. Exp. Med. Biol. Vol. 12, *Morphological and Functional Aspects of Immunity*, K. Lindahl-Kaessling, G. Alm and M. G. Hanna, Jr. (Eds.), Plenum, New York (1971), 323-331.

10. H. N. Eisen, *Immunology* (reprinted from Davis, Dulbecco, Eisen, Ginsberg and Wood, *Microbiology*, 2nd ed.), Harper and Row, New York (1974).

11. M. Feldman, "Cell interactions in the immune response in vitro," in *The Lymphocyte-Structure and Function*, J. J. Marchalonis (Ed.), Marcel Dekker, New York (1977).

12. H. Freedman and J. Gatica, "A threshold model simulating the immune response to replicating antigen," *Math. Biosci.* 37(1977), 113-134.

13. J. Gatica and P. Waltman, "A singular functional differential equation arising in an immunological model," in *Ordinary and Partial Differential Equations*, W. N. Everett and B. D. Sleeman (Eds.), Springer-Verlag, Berlin (1976), 114-124.

14. E. S. Golub, *The Cellular Basis of the Immune Response*, Sinauer, Sunderland, Mass. (1977).

15. J. Hale, *Ordinary Differential Equations*, Wiley-Interscience, New York (1969).

16. J. S. Hege and L. J. Cole, "A mathematical model relating circulating antibody and antibody forming cells," *J. Immunol.* 97(1966), 34-40.

17. G. W. Hoffmann, "A theory of regulation and non-self discrimination in an immune network," *Eur. J. Immunol.* 5(1975), 638-647.

18. G. W. Hoffmann, "Incorporation of nonspecific T-cell dependent helper factor into a network theory of the regulation of the immune response," in *Theoretical Immunology*, G. I. Bell, A. S. Perelson and G. H. Pimbley (Eds.), Marcel Dekker, New York (1978), 571-602.

19. I-D. Hsü and N. D. Kazarinoff, "Existence and stability of periodic solutions of a third order nonlinear autonomous system simulating immune response in animals," *Proc. Roy. Soc., Edin., Ser.* A., 77(1977), 163-175.

20. N. K. Jerne, "The immune system," *Sci. Amer.* 229(1973), 52-60.

21. N. K. Jerne, "The immune system, a web of V domains," *The Harvey Lectures*, Series 70, Academic Press, New York (1976).

22. M. Jilek, "The number of immunologically activated cells after repeated immunization (a mathematical model)," *Folia Microbiologica* 16(1971), 12-23.

23. M. Jilek, "On contacts of immunocompetent cells with antigen (note on a probability model)," *Folia Microbiologica* 16(1971), 83-87.

24. M. Jilek and J. Sterzl, "Modelling of the immune processes," in *Morphological and Functional Aspects of Immunity*, Plenum, New York (1971), 333-349.

25. W. Jilek and J. Sterzl, "On a theory of the immune response," trans. of the *6th Prague Conference on Information Theory, Statistical Decision Functions, Random processes* (Prague 1971), Academia Pub.-Czechoslovak Acad. Sci. 1973, 275-289.

26. M. Jilek and Z. Ursinyova, "The probability of contact between the immunocompetent cell and antigen," *Folia Microbiologica* 15(1970), 294-302.

27. D. D. Joseph and D. H. Sattinger, "Bifurcating time periodic solutions and their stability," *Arch. Rat. Mech. Anal.* 45(1972), 79-109.

28. D. H. Katz, *Lymphocyte Differentiation, Recognition and Regulation*, Academic Press, New York (1977).

29. R. Lefever and R. Garay, "A mathematical model of the immune surveillance against cancer," in *Theoretical Immunology*, G. I. Bell, A. S. Perelson and G. H. Pimbley, Jr. (Eds.), Marcel Dekker, New York (1978), 481-518.

30. J. E. Marsden and M. McCracken, *The Hopf Bifurcation and its Applications*, Springer-Verlag, New York (1976).

31. S. J. Merrill, "A mathematical model of B-cell stimulation and humoral immune response," Ph.D. thesis, U. of Iowa, Dec. 1976.

32. S. J. Merrill, "A geometrical study of B-cell stimulation and humoral immune response," in *Nonlinear Systems and Applications*, V. Laksmikantham (Ed.), (1977), 611-630.

33. S. J. Merrill, "A model of the stimulation of B-cells by replicating antigen," in two parts, *Math. Biosci.* 41(1978), 125-141 and 143-155.

34. S. J. Merrill, "Limit cycles in a model of B-cell stimulation," in Chapter 10 of this volume.

35. R. R. Mohler, C. F. Barton and C-S. Hsü, "T and B cell models in the immune system," Chapter 14 in *Theoretical Immunology*, G. I. Bell, A. S. Perelson and G. H. Pimbley, Jr. (Eds.), Marcel Dekker, New York (1978), 415-436.

36. G. S. V. Nossal, "The cellular and molecular basis of immunological tolerance," in *Essays in Fundamental Immunology*, I. Roitt (Ed.), Blackwell Sci., Oxford, England (1973).

37. A. S. Perelson, "Models of the events responsible for antibody production by B lymphocytes," in *Theoretical Immunology*, G. I. Bell, A. S. Perelson and G. H. Pimbley, Jr. (Eds.), Marcel Dekker, New York (1978), 171-214.

38. A. S. Perelson, M. Mirmirani and G. F. Oster, "Optimal strategies in immunology. I. B-cell differentiation and proliferation," *J. Math. Biol.* 3(1976), 325-367.

39. G. H. Pimbley, Jr., "Periodic solutions of predator-prey equations simulating an immune response," in 2 parts, *Math. Biosci.* 20(1974), 27-51; 21(1974), 251-277.

40. G. H. Pimbley, Jr., "Periodic solutions of third order predator-prey equations simulating an immune response," *Arch. Rat. Mech. Anal.* 55 (1974), 93-123.

41. G. H. Pimbley, Jr., "Bifurcation behavior of periodic solutions of the third order simulated immune response problem," *Arch. Rat. Mech. Anal.* 64(1976), 169-192.

42. G. H. Pimbley, Jr., "Simple time-evolution equations simulating an immune response: their derivation, treatment, and interpretation," in *Theoretical Immunology*, G. I. Bell, A. S. Perelson and G. H. Pimbley, Jr. (Eds.), Marcel Dekker, New York (1978), 539-570.

43. P. H. Richter, "A network theory of the immune system," *Eur. J. Immunol.* 5(1974), 93-123.

44. P. H. Richter, "The network idea and the immune response," in *Theoretical Immunology*, G. I. Bell, A. S. Perelson and G. H. Pimbley, Jr. (Eds.), Marcel Dekker, New York (1978), 539-570.

45. R. Thom, *Structural Stability and Morphogenesis*, W. A. Benjamin, Reading, Massachusetts (1975).

46. P. Waltman, "A threshold model of antigen-stimulated antibody production," Chapter 15 in *Theoretical Immunology*, G. I. Bell, A. S. Perelson and G. H. Pimbley, Jr. (Eds.), Marcel Dekker, New York (1978), 437-454.

47. P. Waltman and E. Butz, "A threshold model of antigen-antibody dynamics," *J. Theor. Biol.* 65(1977), 499-512.

48. J. L. Winkelhake, "A dynamic continuum model for molecular regulation of the humoral immune response," *J. Theor. Biol.* 60(1976), 37-49.

MATHEMATICAL MODELS OF DOSE AND CELL CYCLE EFFECTS IN MULTIFRACTION RADIOTHERAPY

Howard D. Thames, Jr.

Department of Biomathematics
M. D. Anderson Hospital and Tumor Institute
The University of Texas System Cancer Center
Houston, Texas

INTRODUCTION

Two characteristics of treatment method common to most modern radiotherapy centers are the use of very high energy sources for external beam therapy and the division of the total dose applied for tumor control into many small doses ("fractions"; thus the term *multifraction radiotherapy*). High energy beams allow sparing of the skin overlying deep-seated tumors; fractionation of radiation dose decreases damage to skin and other normal tissues affected by the beam. Whereas the underlying mechanism for skin sparing with very high energy beams is primarily *physical*, the sparing of tissues by application of many small doses has been studied mainly in terms of *biological* mechanisms. In this review, phenomena of the latter type will be described which are relevant to the practice of radiotherapy.

The killing of certain numbers of cells by radiation and the consequent alterations in behavior of the survivors are sometimes expressible in terms of mathematical representations, or "models." This possibility stems mainly from the precision with which radiation dose may be specified. To the extent that such a mathematical model faithfully reproduces or simulates the response of the system to a given dose of ionizing radiation, it is possible to simplify the analysis of the response by allowing separation of its different components. Moreover, the construction of models focuses attention on exactly what it is that we do and do not know about the system's response, and thus may guide the design of experiments.

The purpose of this article is to present for the mathematical reader a discussion of radiation response in connection with cell-cycle phenomena and a review of methods used to represent them mathematically. In keeping with the spirit of attempting to understand some of the main features of human response to radiotherapy, emphasis is placed more on "applied" than "theoretical" aspects. It should be understood that the points to be considered are in no sense comprehensive in their potential predictive value for the outcome of radiotherapy. On the other hand, it is to be hoped that some limited, though important, guidelines might be constructed from a successful model of acute and late responses to a course of radiotherapy delivered over several weeks.

THE CELL CYCLE

When it is viewed as a whole, the human body is composed primarily of cells and tissues that proliferate very slowly (tissues that turn over, or replace their original number, in times longer than a few months). This group includes vasculo-connective tissues and parenchymal cells of organs such as liver, kidney, and lung. Because of their slow proliferative rate, these tissues manifest what is termed delayed or "late" response to a course of radiotherapy for control of a human tumor. Late responses appear many months to years following radiotherapy, and are usually considered the limiting factor in determining how large a radiation dose may be delivered for tumor control.

Among the rapidly proliferating (turnover times of the order of one to several days) tissues of the human body are skin, the gastrointestinal tract (mouth to rectum), linings of the nasal air passages, bone marrow, and some tumors. These tissues manifest what is termed acute or early radiation damage, demonstrating more overt sensitivity to radiation than the slowly proliferating tissues, whose response is delayed and often insidious. Rapidly proliferating tissues are usually in a steady-state, with cell production exactly balancing cell loss, and are able to offset their losses following radiation insult by regeneration if sufficient time is allowed, both between treatments and overall, and if the doses given are not of such magnitude as to render recovery impossible.

The content of these remarks is summarized in Table 3.1. From them can be drawn two insights into how the kinetic behavior of cells and their response to radiation influence radiotherapy: (1) The type of radiation damage observed appears to be connected with kinetic factors (thus, the dose-limiting normal tissues may be classified kinetically); (2) Sparing of proliferative normal tissues during radiotherapy depends on both size of treatment dose and time interval between treatments.

TABLE 3.1

Classification of Tissues by Rate of Proliferation

Tissue Type	Tissue	Turnover Time[a]	Radiation Response Type
Proliferative (cell-renewal systems)	Skin G I tract Lining of nasal air passage Bone marrow Tumor	hours to days	Acute (weeks)
Slowly Proliferative (differentiated cells)	Connective Capillary lining Liver Kidney Lung Muscle Bone Cartilage	months	Late (months to years)
Nonproliferative	Nerve	∞	

[a]Time required to replace steady-state population.

Analysis of the proliferative behavior of tissues begins with consideration of the events that occur between cell divisions. These events are collectively called the "cell cycle." The situation is illustrated by Fig. 3.1, where a single cell on the left is shown to produce two daughters, which

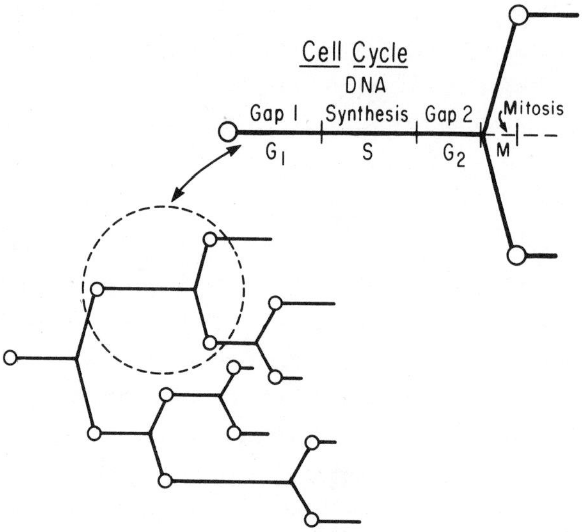

Fig. 3.1. The progeny of a single cell, itself the result of binary fission, with details of interdivisional events shown in the insert. The unequal length of the lines connecting mother cells with points of division to produce daughters demonstrates the variability in cell-cycle times T_c in an otherwise homogeneous population. As shown in the insert, DNA replication (S phase) is separated from cell division (M phase) by gaps 1 and 2 (G_1 and G_2 phases).

divide in their turn to produce two daughters each, etc. The unequal length of the lines which connect an individual cell with the point of its division are intended to illustrate the occurrence in the population of unequal times between the creation of a cell and its division. This life span of a single cell is usually termed the *cell-cycle time* or *generation time*, and is denoted by T_c. The inset shows a breakdown of the events that occur during the time T_c, customarily divided into four phases:

G_1: the gap that occurs between the end of mitosis and the beginning of DNA synthesis;

S: the period during which the cell's DNA is duplicated;

G_2: the gap that occurs between the end of DNA synthesis and the beginning of mitosis;

M: mitosis, whose result is binary fission of the cell.

In classical cell biology only the broad classification of the stages of mitosis (prophase, metaphase, anaphase, telophase) was possible, because of their visibility through the microscope. The long period (relative to duration of mitosis) between mitoses was referred to as interphase, and consisted of what are now called G_1, S, and G_2. The discrimination between G_1, S, and G_2 has been made possible by techniques developed during the last 25 years [30].

Determination of Phase-transit Times

Experimental determination of T_c is based on a technique known as autoradiography [42], and more recently pulse cytophotometry [41]. The autoradiographic technique consists in administering a radioactive DNA precursor (^{3}H-Tdr) for a period small relative to T_c. Cells in S-phase are labeled, and their progress through subsequent divisions may be followed by measuring the fraction of labeled mitoses on photographic emulsions. By analyzing the percent labeled mitoses (PLM) data, the durations of the phases of the cell cycle may be estimated. The method is explained in Fig. 3.2a. As shown there, the ^{3}H-Tdr administered at t = 0 is manifested in labeled mitoses after a time corresponding roughly to T_{G_2}, the G_2-transit time. The first labeled cohort passes through mitosis, and its daughters cause a second peak T_c hours later. Experimental results are usually more akin to the curve in Fig. 3.2b because of the variation in values of T_c in most populations of cells.

The pulse cytophotometric (PCP; also "flowmicrofluorometric") technique allows the measurement of the DNA content per cell in a cell population, and the display of the frequency distribution of this value, usually in the form of a histogram. As shown in Fig. 3.3, sharp peaks mark those cells in the population that are either pre- or post synthesis (G_1 and G_2 cells). The application of this technique to the determination of cell-cycle parameters is still in a developmental state [7, 19] and will not be discussed further here.

In addition to estimates of the generation time T_c, labeling techniques provide useful information concerning the fraction of cells of a population that is in S-phase, and the fraction of cells that is in cycle. Thus, the labeling index (LI) = fraction of cells incorporating ^{3}H-Tdr after a pulse of short duration = fraction of cells in S-phase. The continuous labeling index (CLI) is defined as the fraction of cells labeled from a precursor pool of tritiated thymidine with constant specific activity. It can often

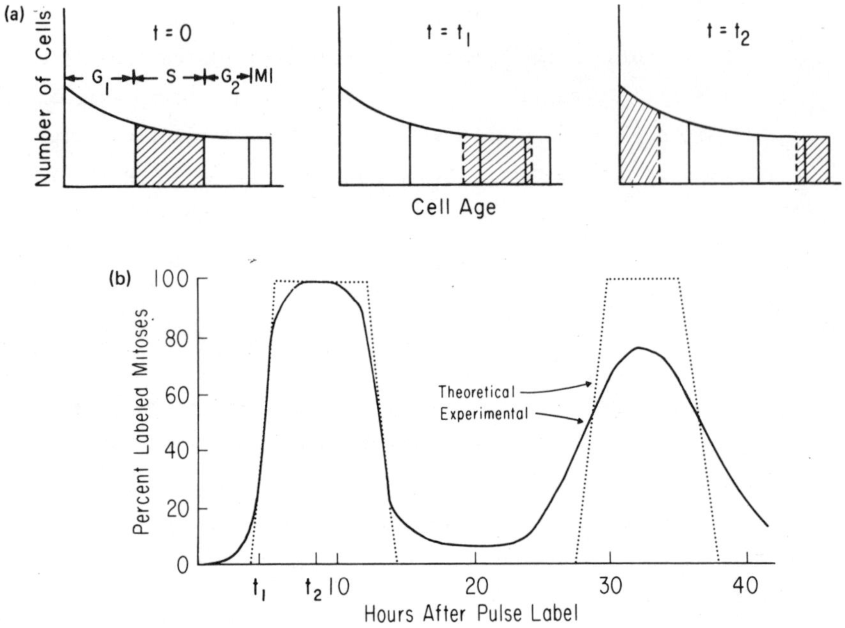

Fig. 3.2. Determination of Percent Labeled Mitoses (PLM).
a. At $t_o = 0$, the portion of cells present in S phase is radioactively labeled by addition of ^{3}H-Tdr for a time short compared with T_c. At $t = t_1$ the labeled cohort has aged, so that its leading edge is in mitosis. On the associated curve (Fig. 3.2b) a small nonzero value of PLM is recorded. At $t = t_2$, all mitotic cells are labeled (PLM = 100%). The second peak is recorded when the labeled cohort traverses M phase the second time. The distance between peaks is an estimate of T_c.
b. Experimentally observed PLM curve. Loss of synchrony caused by variation in cell-cycle transit times results in loss of the sharpness shown in the hypothetical PLM curve (dotted overlay).

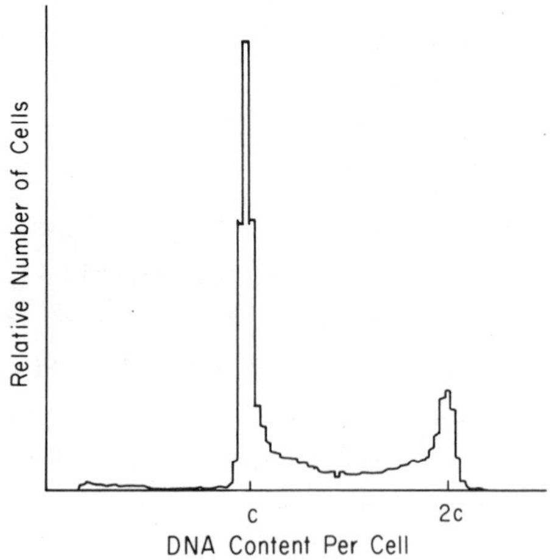

Fig. 3.3. Distribution of cells with a given DNA content per cell determined by pulse cytophotometry (PCP). Peaks at c (G_1) and 2c (G_2) correspond to pre- and post-synthesis DNA content.

be deduced from the rate at which CLI → 1 whether a significant fraction of noncycling cells is present. Finally, the mitotic index (MI), or fraction of mitotic cells, is a measure of T_M/T_c in populations in which nearly all cells are cycling (as the LI is a measure of T_S/T_c), where T_S and T_M represent, respectively, the transit times through S phase and mitosis.

A shortcoming of the autoradiographic methods of cell kinetic analysis is that results are biased in favor of rapidly cycling cells, and moreover apply only to the proliferative compartment. Pulse cytophotometric techniques allow much more rapid determinations of cell kinetic parameters, and the recent development of two-dimensional PCP with differentially staining DNA dyes promises to allow discrimination between cycling and noncycling cells [52]. In examining irradiated populations, both of the above methods have the serious disadvantage of being unable to resolve reproductively dead but morphologically intact cells from surviving cells with infinite reproductive potential (clonogenic cells).

Mathematical Models (Nondynamic)

Mathematical models of cell-cycle kinetics can be divided into two groups: nondynamic and dynamic. Nondynamic models are used in the analysis of experimental data from cell populations growing under stationary, or steady-state, conditions. The word stationary refers to the invariance of the cell population age distribution; the total number of cells increases exponentially (thus also the terms exponential and "log-phase" growth). Dynamic models are required to analyze data from populations where transient variations occur. Examples of such variations are perturbations introduced into the age distribution with radiation or drugs, diurnal variations in cycle parameters, and decreased growth rate (increased mean T_c) resulting from the approach of the population density to saturation conditions.

In the following, different stationary models of the cell-cycle will be divided into three somewhat arbitrary classes. Emphasis will be placed on a class which we label discrete state-vector models, a choice motivated by two considerations. First, these models have been extended to the dynamic setting more often than those of the other groups. Second, they fit more naturally into modeling situations in which radiation is the cause of perturbations in the age distributions.

1. Branching-process models [5, 23, 37, 38, 57]. The proliferating cell population is represented as an age-dependent branching process in

which the lifetime of each cell is the sum of four phase durations:

$$T_c = T_{G_1} + T_S + T_{G_2} + T_M$$

and for which the mean number of daughter cells per division is $Q \in [1, 2]$. If $\phi(t)$ is the probability density of T_c, and $L(s)$ is the Laplace transform of ϕ, the rate of increase of the population is the positive root k of the equation

$$QL(k) = 1 \tag{1}$$

and the asymptotic density of cells aged between [t, t + dt] is [24]

$$g(t) = \frac{Q}{Q-1} ke^{-kt}\left[1 - \int_0^t \phi(\tau)d\tau\right] \tag{2}$$

Thus the proportion of the expected population size in any specified phase or age group approaches a limit that is independent of time. Assuming that cells have this "stable" age distribution in the cycle, it is possible to simulate theoretically the PLM curve.

The best known practical fitting method based on branching-process theory is that of G. G. Steel [49] and J. C. Barrett [4]. Suppose the rancom variables T_{G_1}, T_{G_2}, and T_S have probability density functions (p.d.f.) $f_1(t)$, $f_2(t)$, and $f_3(t)$, respectively (we assume $T_M = 0$). The corresponding p.d.f.s for durations of $G_2 + S$, $G_2 + S + G_1$, $G_2 + S + G_1 + G_2$, $G_2 + S + G_1 + G_2 + S$, etc., are $h_1(t)$, $h_2(t)$, $h_3(t)$, and $h_4(t)$, respectively, where

$$\begin{aligned} h_1(t) &= \int_0^t f_3(t - x)f_2(x)dx \\ h_{i+1}(t) &= \int_0^t h_i(t - x)f_i(x)dx, \qquad i = 1, 2, \ldots \end{aligned} \tag{3}$$

Let $h_0(t) = f_2(t)$ and define the cumulative distributions

$$H_i(t) = \int_0^t h_i(t')dt', \qquad i = 0, 1, 2, \ldots$$

Then $H_1(t) = \text{prob}\ \{T_{G_2} + T_S \leq t\}$ and $H_0(t) = \text{prob}\ \{T_{G_2} \leq t\}$ so that

$$H_0(t) - H_1(t) = \text{prob}\ \{T_{G_2} \leq t \leq T_{G_2} + T_S\}$$

$$= \text{prob}\ \{\text{a cell mitotic at time } t \text{ was in S phase at time of labeling}\}$$

It is also possible that a mitotic cell is labeled whose 1st, 2nd, etc. ancestor was in S-phase at the time of labeling, whence

$$H(t) = H_0(t) - H_1(t) + H_3(t) - H_4(t) + \ldots$$

gives the fraction of labeled mitoses.

At this point the Laplace transforms of the convolutions in Eq. (3) are obtained, resulting in

$$p\,H^*(p) = \frac{f_2{*}(p)\ (1 - f_3{*}(p))}{1 - f_1{*}(p)\ f_2{*}(p)\ f_3{*}(p)}$$

where * denotes Laplace transform. The method developed by Barrett employs Monte Carlo techniques to generate phase durations from specified distributions, whose parameters are adjusted to agree with experimental PLM data, instead of using Laplace transforms as described above to generate "analytical" results. The analytical approach is taken by Trucco and Brockwell [57], MacDonald [37], and Hartmann and Pedersen [25]. The details may be found in the cited references.

2. Continuous maturity-time models [44, 51, 58]. In these models the dependent variable is the population density, which depends on time and cell maturity. The latter is a measure of various stages of growth in the cell, such as birth, onset of DNA synthesis, onset of mitosis, etc. Thus maturity, or physiological age, is often distinct from chronological age, in whose terms branching-process models are defined. For example, in the latter models the time at which mitosis of labeled cells occurs is specified probabilistically, as a random variable. In contrast, mitosis always occurs at the same physiological age in maturity-time models.

The equation derived by von Foerster [58] for cell density n(x, t) is

$$\frac{\partial n}{\partial t} + \frac{\partial n}{\partial x} = -\lambda n$$

with the boundary condition

$$n(0, t) = Qn(1, t)$$

Rubinow [44] generalized von Foerster's equations by allowing a variable maturation velocity $v = v(x, t)$:

$$\frac{\partial n}{\partial t} + \frac{\partial}{\partial x}(vn) = -\lambda n$$

Variability in generation times is assumed to arise from differing maturation velocities, distributed according to an assumed form.

Weiss [60] derives Markovian equations that generalize those of von Foerster. Thus

$$\frac{\partial n}{\partial t} + \frac{\partial n}{\partial x} = -(\lambda + \Phi + \rho)n + \int_0^a n(a, x - x', t)\psi(a, x - x', x', t)dx'$$

in which a = chronological age and x = maturity. $\psi(a, x, x', t)$ is the rate of maturation in a time interval dt. The Fokker-Planck approximation to this equation (in which it is assumed that the increment in physiological age is small in the time interval (t, t + dt)) leads to the Stuart-Merkle [51] model

$$\frac{\partial n}{\partial t} = -v_0 \frac{\partial n}{\partial x} + D \frac{\partial^2 n}{\partial x^2} \tag{4}$$

with $Qn(1, t) = n(0, t)$ as above. Here D is a "diffusion" constant, and v_0 a "convection" velocity. Variability occurs in the maturation velocity by a random walk process in the variable x; the mean maturation velocity is v_0. Thus the generation time is different from one cell to another, although mitosis always takes place at the fixed value of maturity $x = 1$.

The asymptotic behavior of solutions of Eq. (4) is given by (Hethcote and Thames, unpublished)

$$n(x, t) \to N(x) = Q^{-x} e^{kt} \int_0^1 Q^z n(z, 0)dz$$

$$k = ((1 + 2(D/v_0^2)\ell n\, Q)^2 - 1)/4(D/v_0^2)$$

N(x) has a time-independent, exponential form similar to that of the asymptotic cell density g(t) (Eq. (2)) from branching-process models.

A disadvantage of the model is that it predicts backward movement in the cycle, inasmuch as generation times are approximately normally distributed about $1/v_0$. The principal advantage of the model, as with the other maturity-time models, is that physiological-age-dependent effects of radiation and drugs (customarily abbreviated to "age response to radiation and drugs") can be modeled more naturally than with branching-process models. One simply modifies n(x, t) by a fraction representing the observed survival response of the population at age x. This advantage, along with considerable mathematical simplification and intuitive appeal, is shared by the so-called state-vector models we describe next.

3. Discrete maturity-time (state-vector) models [21, 34, 53]. For these models the spectrum of maturity levels (usually taken as the unit interval in the continuous models discussed above) is discretized into k nonoverlapping, independent compartments. The density of cells at maturity level x_i, $i = 1, \ldots, k$, is then given by the element y_i of the population state vector $y = (y_1, \ldots, y_k)$. Further subdivision of k in the form

$$k = k_{G_1} + k_S + k_{G_2} + k_M$$

is sometimes convenient, e.g., in the fitting of PLM data.

The evolution of the state vector y(t) is defined in the Takahashi [53] model by

$$\frac{dy}{dt} = Ay$$

where

$$A = \begin{bmatrix} -\lambda & & & & Q\lambda \\ \lambda & -\lambda & & & \\ & \lambda & -\lambda & & \\ & & & \ddots & \\ & & & \lambda & -\lambda \end{bmatrix} \qquad (5)$$

This model is a special case of a multiple-phase birth process ([34]

Eq. (17)). Transit times in each compartment are exponentially distributed, and the generation time T_c has the gamma distribution. λ is defined by

$$\lambda = \frac{k}{E(T_c)} \tag{6}$$

The variance of T_c can only assume discrete values since

$$\mathrm{var}\,\frac{T_c}{E(T_c)^2} = \frac{1}{k} \tag{7}$$

The asymptotic fraction of the total number of cells in the ith compartment is given by [32]

$$\frac{y_i(\infty)}{\sum_{i=1}^{k} y_i(\infty)} = \frac{Q}{Q-1}(Q^{1/k} - 1)\, Q^{-i/k}, \qquad i = 1, \ldots, k$$

The Hahn [21] model describes the progression of cells through the cycle as a discrete-time Markov process:

$$y(t + \Delta) = By(t) \tag{8}$$

In Eq. (8) Δ is a small (relative to T_c) time increment and

$$B = \begin{bmatrix} \alpha & & & & \gamma Q & \beta Q \\ \beta & \alpha & & & & \gamma Q \\ \gamma & \beta & & & & \\ & \gamma & & & & \\ & & \ddots & & & \\ & & & \alpha & & \\ & & & \beta & \alpha & \\ & & & \gamma & \beta & \alpha \end{bmatrix} \tag{9}$$

Here β = prob (cell of age i will advance to age i + 1 during Δ), γ = prob (cell of age i will advance to age i + 2 during Δ), and $\alpha = 1 - \beta - \gamma$ is the probability that the cell will not advance in age during time Δ. The rank of B is designated k. The sequence of vectors $By(0)$, $B^2y(0), \ldots$ tends asymptotically to the eigenvector of B corresponding to the largest eigenvalue

$$\lambda = \alpha + \beta Q^{1/k} + \gamma Q^{2/k} \tag{10}$$

The stable age distribution has the same form as that presented above for the Takahashi model, i.e., the relative number of cells aged i is proportional to $Q^{-i/k}$, $i = 1, \ldots, k$.

The requirements $\alpha, \beta, \gamma \in [0, 1]$ lead to certain constraints on allowable k's for a Hahn model [55]. These authors show that the transit time of the mean of the age p.d.f. of an initially synchronized cohort, T_M is given by

$$T_M = \frac{k\Delta}{\beta + 2\gamma} \tag{11}$$

and its variance σ_M^2 by

$$\sigma_M^2 = T_M\Delta\left(\frac{\beta + 4\gamma}{(\beta + 2\gamma)^2} - 1\right)$$

Then k's for which α, β, and γ lie in the unit interval are in the range

$$\frac{T_M^2}{\sigma_M^2 + T_M\Delta} \leq k \leq \frac{2T_M^2}{\sigma_M^2 + T_M\Delta}$$

The mean generation time for a cell population described by the Hahn model is given by [69]

$$T_c = \frac{k}{\beta + 2\gamma} + \frac{\gamma}{(\beta + 2\gamma)^2}\left(1 - \left(\frac{-\gamma}{\beta + \gamma}\right)^k\right)$$

The variance of the generation time σ^2 is computed numerically.

The discreteness of the variance of the generation times in the Takahashi model (Eq. (7)) leads to numerical difficulties in fitting PLM data. They may be avoided by fitting the data first to a Hahn model [55], and using the resulting mean and variance of the transit times to define the parameters of a "closest" Takahashi model (Eqs. (6), (7)).

MATHEMATICAL MODELS (DYNAMIC)

The discrete state-vector models discussed above have been most widely used in simulating the proliferative characteristics of populations that have time-varying mean generation time, nonstable age structure, or other qualities which render their modeling by nondynamic models unsuitable. That varying mean generation times may not be of limited occurrence is illustrated by the data of Collyn-d'Hooghe et al. [9]. The average cycle and G_1 durations vary significantly according to generation in a cultured cell population: a decrease during the first three generations after monolayer culture, and then an increase with the age of the culture. By contrast, mitotic times remain relatively constant.

Jansson [31] developed a dynamic model of the Takahashi-type (Eq. (5)) in which flow rate λ through the cycle decreases with increasing cell population:

$$\lambda(t) = \frac{\lambda_0}{1 + \delta \sum_{i=1}^{k} y_i(t)} \tag{11a}$$

The parameters λ_0 and δ determined from fits to PLM data (for 2-, 6-, and 10-day ascites tumors) were used to generate a theoretical growth curve. The later was in excellent agreement with experiment, providing a consistency check of the method (cf. Figures 5 and 7, op. cit.).

The Hahn model has been applied to the study of radiation [22, 40] and drug [43] effects on cell proliferation. These applications involve some amount of data from the modeling of dose-response curves to be discussed in the next section. Their treatment will be postponed for inclusion in section V.

Finally, Klein and Valleron [35] have included the effects of diurnal rhythms in cell kinetics, using a Takahashi model (Eq. (5)) with periodic flow rates through G_1 to S. If the latter has Fourier series

$$u(t) = u_0 + \sum_{k=1}^{\infty} u_k \sin(kwt + \phi)$$

then the labeling (LI(t)) and mitotic (MI(t)) indices are periodic with Fourier series

$$LI(t) = a_0 + \sum a_k \sin(kwt + \phi + \psi)$$

$$MI(t) = b_0 + \sum b_k \sin(kwt + \phi + \eta)$$

The authors show that

$$\frac{a_0}{b_0} = \frac{T_s}{T_M}$$

$$\frac{a_k}{b_k} = S_k(T_s) \cdot G_{2_k}(T_{G_2}) \cdot M_k(T_M)$$

where each factor is dependent only on parameters from the indicated phase, and that the time lag $\psi - \eta$ between equivalent points on the LI(t) and MI(t) curves are given by

$$\psi - \eta = s\ (T_S) + g_2\ (T_{G_2}) + m\ (T_M)$$

i.e., the time lag is a sum of functions whose arguments are particular to one of the phases S, G_2, or M.

Cell Production and Cell Loss

Using autoradiographic methods, Hermens [26] addressed the question of whether the growth of a tumor could properly be explained by the increase in size of a pure p-(proliferative) population, or by that of a population consisting of a mixture of p- and q-(quiescent) cells. The latter are cells which have left the proliferative pool. He found that q cells are present at all levels of tumor growth investigated, and that the fraction of q cells increases with increasing tumor size. The presence of these cells indicates that the average number of p cells produced per dividing cell must be less than 2; i.e., $Q < 2$. In fact, the data for a rhabdomyosarcoma show that $Q \simeq 1.5$ for small (1 cm^3) tumors, and decreases to values in the range 1.2 to 1.3 for larger tumors (1500 to 3000 cm^3), depending on location in the center or at the periphery of the tumor.

The following deterministic model was introduced in an attempt to account for *cell production* and *cell loss* in tumors [26, 46, 47]. Observed parameters are the volume doubling time T_d and the rate constant for growth, k. These correspond to the model parameters potential doubling time $T_{d(pot)}$ and rate constant for the production of new p cells, k_{prod}. It is found that

$$k_{prod} = \left(1 - \frac{R}{Q}\right) \frac{\ln Q}{T_c}$$

and

$$T_{d(pot)} = \frac{\ln 2}{k_{prod}} = \left(\frac{Q}{Q - R}\right) \frac{\ln 2}{\ln Q} T_c$$

In these equations R represents the fraction of p cells which, at mitosis, will not be able to divide successfully. It is determined experimentally as the fraction of aberrant mitoses. In the work of Hermens [26], 20 to 28% of cycling cells produce abortive mitoses throughout tumor growth.

Let k_L denote the rate constant for cell loss from the tumor. Such loss may occur by any of several mechanisms [47], including cell lysis and emigration from the tumor. Then the net production per cell at time t is given by

$$\frac{\frac{dn}{dt}}{n} = k_{prod} - k_L$$

or

$$\frac{\ln 2}{T_d} = \frac{\ln 2}{T_{d(pot)}} - k_L$$

from which

$$k_L = \frac{\ln 2}{T_{d(pot)}} \left(1 - \frac{T_{d(pot)}}{T_d}\right)$$

$$\equiv \phi \frac{\ln 2}{T_{d(pot)}}$$

Here ϕ is Steel's "cell-loss factor."

Hermens [26] found that the potential doubling time predicted for the larger tumors (1500 to 3000 cm^3) was smaller than that measured:

$$T_{d(pot)} = 60 \text{ hrs} < 88 \text{ hrs} = T_d \tag{12}$$

Thus, cell loss would appear to play a major role in governing the growth of solid tumors. A summary of some of Hermens' results for rhabdomyosarcoma is presented in Table 3.2. There it is evident that macroscopic growth is characterized by a decrease in growth rate with increase in tumor volume, attributable possibly to both a decreasing rate of cell production and an increasing rate of cell loss. However, while wide variation in generation times was observed for different locations within the tumor, there is no significant change in T_c with increasing tumor size. Thus, the decreased growth rate cannot be explained by an increase in the mean generation time.

TABLE 3.2

Cell Loss and Production Associated with Growth of an Experimental Rhabdomyosarcoma

Tumor Volume (cm^3)	T_c (hr)	T_d (hr)	$T_{d(pot)}$ (hr)	Q	R	Growth Fraction[a]	k_{prod} (cells hr^{-1}/cell)	k_L (cells hr^{-1}/cell)	ϕ
1 to 5	19.6	60±10	41.4	1.485	.253	.37	.0167	0	.31±.12
300	20.5	50.4	50.7	1.410	.261	.30	.0137	0	0
1500 to 3000	20.5	87.6	59.7	1.322	.195	.26	.0116	.0037	.32

[a]The fraction of in-cycle cells (from Hermens, (1973)).

To summarize, tumors must not be regarded simply as expanding cell populations, whose growth can be stopped only by interference with cell production, but rather as cell renewal systems which are slightly out of balance--systems in which the rate of cell production only slightly exceeds the rate of cell loss [48]. The overall growth of the tumor, which clinically is what is of most concern, is critically dependent on the competition between cell production and cell loss.

RADIATION DAMAGE

The most precise measurements of radiation effects on cells result from the single-cell survival assay, in which "killing" is interpreted as loss of unlimited reproductive capability. N_i cells are exposed to doses x_i ($i = 0, \ldots, k$; $x_0 = 0$) of radiation, and plated onto nutrient-agar suspension. Cells which retain reproductive (or clonogenic) capability produce colonies, which may be counted. Assuming that each colony results from one surviving cell (a multiplicity correction can be made), and that n_i colonies result from exposure of N_i cells to dose x_i,

$$\text{surviving fraction } (x_i) = \frac{\dfrac{n_i}{N_i}}{\dfrac{n_o}{N_o}}, \qquad i = 0, \ldots, k$$

Cell-survival curves may only be established for cells that are in cycle; less precise methods must be resorted to for cells described in Section II as very slowly proliferating (see "*in vivo* data" below).

Cell-Survival Models

Because large numbers of cells (10^5) may be prepared for use in survival assays, it is possible to measure survival over several decades. Thus it is natural to work with a semilogarithmic scale. Typical of the curves that result are those shown in Fig. 3.4. Their dependence on dose may be interpreted in terms of *target-theory* models, which, although no longer felt to be accurate biophysically, are useful conceptually and will therefore be described.

Suppose that, as a result of the absorption of energy from the radiation beam, a density ξ of potentially damaging events exists, these caused primarily by fast-moving electrons set into motion by Compton interactions between X-rays and the biological material in the cell. Assume further

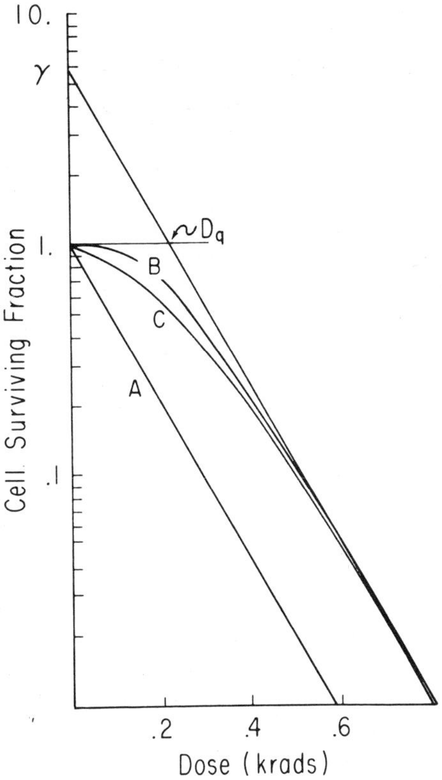

Fig. 3.4. Single-dose cell-survival curves deduced from target theory (see text). Curves A (single-hit, or exponential, survival), B (multi-target, single-hit) and C ("two-component" product of curves A and B). γ, called the *extrapolation number* (theoretically, the number of targets in curves B and C), is obtained by extrapolating the asymptote of the survival curve to the survival axis. The asymptote has slope $-(\alpha + \beta) \equiv -1/D_o$. Hence D_o is an *inverse measure of radiation sensitivity* at high (usually nonclinical) doses. The intercept of the asymptote with the dose axis is the *quasi-threshold dose* D_q. $D_q = D_o \ln \gamma$ is a measure of the "shoulder" width (zero for curve A) of threshold-type survival curves (B and C).

that each cell of the population presents a volume cross-section σ for biologically damaging interaction; σ is the "target size" of the cell. If the total cell volume of the population is V, then

$$\rho \equiv \text{prob}\ \{\text{a damaging event registers in a cell}\}$$

$$= \frac{\sigma\xi}{V\xi} = \frac{\sigma}{V}$$

ρ is known as the *hit probability* of a cell. Since $\sigma \leq$ volume of a single cell, ρ is quite small:

$$\rho \leq N^{-1} = 0(10^{-5})$$

The probability that a cell will receive h "hits" from a total of $V\xi$ "tosses" is

$$\binom{V\xi}{h}\rho^h(1 - \rho)^{V\xi-h} = p(\rho, V\xi, h)$$

If a *single hit* suffices to kill the cell, then the surviving fraction is given by

$$p(\rho, V\xi, 0) = (1 - \rho)^{V\xi} \cong e^{-\rho V\xi}$$

which accounts for the pure-exponential survival curve A shown in Fig. 3.4 (ξ is proportional to dose).

Now suppose that each cell has n targets, each with equal probability of being hit, and that a single hit is sufficient to inactivate each target. Carrying over the notation from above, we assume that

$$e^{-\rho V\xi} = \text{prob}\ \{\text{a target is not hit during } V\xi \text{ active events}\}$$

Then the probability that a cell survives with ν targets hit and $n - \nu$ targets missed is

$$P(\rho V\xi, n, \nu) = \binom{n}{\nu}(1 - e^{-\rho V\xi})^n(e^{-\rho V\xi})^{n-\nu}Q(\nu)$$

in which $Q(\nu)$ = prob {cell survives with ν targets hit}. If inactivation of all n targets is required, then the surviving fraction is given by

$$\sum_{\nu=0}^{n-1} \binom{n}{\nu} (1 - e^{-\rho V\xi})^{\nu} (e^{-\rho V\xi})^{n-\nu} = 1 - (1 - e^{-\rho V\xi})^{n}$$

which is curve B shown in Fig. 3.4. This is known as the *multitarget, single-hit* (sometimes simply multitarget) model.

The above derivations have been greatly oversimplified. More detail may be found in Elkind and Whitmore [14].

Curve C of Fig. 3.4 is the product of a curve of type A and a curve of type B. The rationale behind this form is that many radiation beams are heterogeneous in regard to the amount of energy deposited by secondary particles (electrons) along their tracks, defined by the LET (linear energy transfer) distribution for a particular energy in the spectrum of the radiation beam. X-rays and ^{60}Co γ-rays are usually termed low LET beams; neutrons give rise to tracks of high LET. Exponential killing (curve A) characterizes high LET beams, whereas curves with a "shoulder" like that of B are typical of low LET irradiation. It is natural to suppose that a beam with components of both types would lead to a curve of type C, where surviving fraction is given by

$$e^{-\rho' V\xi'} [1 - (1 - e^{-\rho'' V\xi''})^{n}]$$

The primes are used to indicate the different target and dose-rate characteristics of the beam components. This model has been called the "two-component" model. See Elkind [13] for more detail.

In order to simplify the notation, we now summarize the characteristics of the function

$$f(x) = \ell n \text{ (surviving fraction)}$$

for situations similar to those shown in Fig. 3.4. We replace $\rho V\xi$ by αx or βx, where x is the measured absorbed dose and α and β have units rad^{-1}. We introduce a third model ("linear-quadratic") which has no target-theory interpretation, but was developed from the notion of "dual radiation damage" (see [33]).

It is typical of many of these curves that

$$\begin{aligned} &1.\quad f'(x) < 0 \text{ for } x \geq 0 \\ &2.\quad f'(x) \to \text{constant for } x \to \infty \\ &3.\quad f''(x) \leq 0 \text{ for } x \geq 0 \end{aligned} \tag{13}$$

although there is controversy concerning the second point.

Some commonly used mathematical representations of curves similar to f(x) in Figure 3.4 follow.

1. Multitarget model.

$$f_{MT}(x;\ \beta,\ n) = \ell n[1 - (1 - e^{-\beta x})^n] \tag{14}$$

Although $f'_{MT}(0;\ \beta,\ n) = 0$, requirements (2) and (3) are satisfied. In particular, f_{MT} is asymptotic to a line with slope $-\beta$:

$$f'_{MT} \to -\beta \equiv \frac{-1}{D_o} \quad \text{as } x \to \infty \tag{15}$$

where the commonly used symbol "D_o" (inactivation dose) is an inverse measure of the high-dose sensitivity of the population to radiation. n is called the "extrapolation" number. The initial, nonexponential portion of the curve is called its "shoulder." A measure of the width of the shoulder (and of the population's resistance to low doses) is the x-intercept of the high-dose asymptote, which is called "D_q":

$$D_q \equiv \beta^{-1}\ \ell n\ n = D_o\ \ell n\ n \tag{16}$$

2. Two-component model.

$$f_{TC}(x;\ \alpha,\ \beta,\ n) = -\alpha x + \ell n[1 - (1 - e^{-\beta x})^n] \tag{17}$$

Thus $f_{MT} = f_{TC}(x;\ 0,\ \beta,\ n)$. The quantities α and β are usually interpreted as reciprocal mean inactivation doses

$${}_1D_o = \alpha^{-1}, \qquad {}_nD_o = \beta^{-1}$$

and n as the target multiplicity (see Elkind [13] for details). The two-component curve is asymptotic to lines with slope $-\alpha$ (as $x \to 0$) and $-(\alpha + \beta)$ (as $x \to \infty$). The α term is presumed due to a contribution to lethality from single-hit inactivation resulting from a high-LET proportion of the x-ray or γ-ray dose. A majority of survival curves in the literature are characterized by nonzero initial slope.

3. Linear-quadratic model.

$$f_{LQ}(x; \alpha, \beta) = -\alpha x - \beta x^2 \tag{18}$$

A clear advantage of this model is its linearity in the parameters α and β. The $-\alpha x$ term is usually interpreted as the ln [prob (two targets are affected by a single particle)] [33] and is biophysically different from the same term in f_{TC}. f_{LQ} is not asymptotic to a line as $x \to \infty$; however, there is some doubt whether the requirement (2) (Eq. (13)) applies to all survival data, or whether, for some data at least, the survival curve continuously "bends downward" at high doses [62].

Multifraction Survival Data

Conventional radiotherapy uses multiple fractional doses given over an extended time period. These doses are small enough that the effects they may be expected to produce lie predominantly on the "shoulders" of the survival curves discussed in the previous section. The negative curvature in the clinical dose region of the survival curves discussed above has the implication for multifraction irradiation that, within a limit, the total dose required to produce a certain level of injury increases as the number of dose fractions is increased. The "limit" alluded to, both its existence and the estimation of its value, is a matter of critical importance in radiotherapy, and will be discussed below.

The increase in total dose to achieve a certain effect with number of fractions may be understood by referring to Figure 3.5. Suppose that a total dose $2x_2$ given in two fractions (treatments) of x_2 units each is required to inflict the same injury as x_1 units given in a single fraction:

$$2f(x_2) = f(x_1)$$

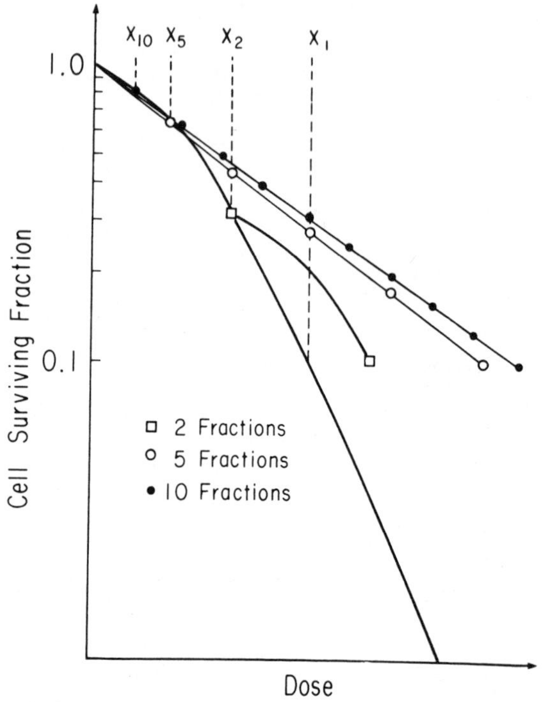

Fig. 3.5. Multifraction cell-survival curves. When sufficient time (approximately 5 hours) elapses between equal-dose treatments, the initial portion of the survival curve repeats itself (cell repopulation is ignored). This, and the negative curvature of the single-dose survival curve, have as a consequence that the total dose necessary for fixed effect increases with increasing number of "fractions" (treatments). Reduction of surviving fraction to 10^{-1} is effected by total doses $10x_{10} > 5x_5 > 2x_2 > x_1$, yet the opposite relationship holds between the doses per fraction: $x_{10} < x_5 < x_2 < x_1$. The n-fraction response curve is exponential with slope $-\alpha_{eff} = -(\alpha + \beta x_n)$ (LQ model, Eq. (18)) where $-\alpha_{eff} = -1/D_{o(eff)}$. The "flexure" dose d_f is that dose per fraction which, given n times, produces an effect that cannot be statistically distinguished from the effect produced by n + 1 fractions of $nd_f/(n + 1)$ rad per fraction. For the curves shown, $n > 5$.

where $f(x) = \ln$ (surviving fraction at dose x). Because of the convexity of survival functions (property (3), Eq. (13)),

$$f(2x_2) \leq 2f(x_2) = f(x_1)$$

In view of this, the monotone decreasing nature of these curves (property (1), Eq. (13)) implies

$$2x_2 \geq x_1$$

with equality only in case the survival curve is linear ($f'' \equiv 0$). The same argument shows that if $mf(x_m) = (m - 1) f(x_{m-1})$, then $mx_m \geq (m - 1)x_{m-1}$, i.e., the total dose necessary for fixed effect increases with increasing number of dose fractions.

As is evident in Figure 3.5, the multifraction response curve approximates more and more closely an exponential in dose as m (the number of dose fractions) increases. Thus, the slope $-\alpha_{eff}$ of the multifraction survival curve is approximately constant, and may be written in terms of the "effective" inactivation dose $D_{o(eff)}$ (see Eq. (15)):

$$-\alpha_{eff} = \frac{-1}{D_{o(eff)}}$$

The effective slope of the multifraction survival curve may be bounded in terms of the parameters of the models discussed above. We find

$\alpha_{eff} \in (0, \beta)$	(Multitarget model, Eq. (14))
$\alpha_{eff} \in (\alpha, \alpha + \beta)$	(Two-component model, Eq. (17))
$\alpha_{eff} = \alpha + \beta x_n \in (\alpha, \infty)$	(Linear-quadratic model, Eq. (18); x_n = dose per fraction, n = number of fractions = 1, 2,...)

Thus, when the dose per fraction is small enough, the slope of the multifraction response curve approaches the initial slope of the single-dose survival curve, which is nonzero for the majority of mammalian cell and tissue dose-response curves available in the literature. The models (14), (17), and (18) have been compared for different types of multifraction data by Hethcote, McLarty, and Thames [28].

As the number of fractions into which the fixed dose x_1 is divided increases, the resulting effect decreases.* The difference between the dose $n\,x_n$ required to produce a given effect E in n treatments of x_n units each and the single dose x_1 which results in effect E is called the *spared dose* d_n:

$$d_n = n\,x_n - x_1 \qquad \text{(effect level E)}$$

It is clear (fig. 3.5) that d_n tends to an upper limit as the number of fractions tends to infinity (i.e., as dose per fraction tends to zero). For example, if effect is defined by the linear-quadratic model Eq. (18), we have

$$\begin{aligned} d_\infty &= \lim_{n\to\infty} d_n \\ &= \frac{E}{\alpha} - x_1 \\ &= \left(\frac{\beta}{\alpha}\right) x_1^2 \end{aligned}$$

Here we made use of the fact that the multifraction response tends to the line with slope $-\alpha$ (cf. Fig. 3.5) as $n \to \infty$, so that the maximum dose for effect E is given by E/α.

The sparing effect of dose fractionation is usually considered to result from the *repair of sublethal injury*. Sublethal injury may be understood (heuristically) by reference to the multitarget theory of radiation damage presented above. At doses per fraction in the "shoulder" region of the survival curve, a proportion of cells sustains hits to less than the n targets required for lethality. These are repaired in a few hours, so that survivors respond to an ensuing dose as though no previous doses had been

*The regeneration of normal cell-renewal systems is also made possible by protracting treatment into many small fractions (discussed in the following section). The sparing discussed above refers primarily to alleviation of late effects in non-proliferative tissues.

delivered, i.e., the shoulder of the curve is repeated. As the dose per fraction is increased, surviving cells become "saturated" with damage; for example, as the fraction of survivors decreases below 0.1, n - 1 targets will very likely be inactivated in those cells which survive ([14] pp. 20-28). This is consistent with the fact that d_∞ increases with increases in the single dose x_1 which causes effect E.

It has been assumed up to now that in a course of n fractions the size of each of the doses (x_n) is the same. This need not be the case, however, and it is commonplace to find altered doses per fraction in the course of treatment of a patient with external beam radiation. In view of this it is of some interest to note that *the spared dose d_n is maximized by using equal doses per fraction x_n* (neglecting the effects of proliferation of the survivors, which in any event do not apply to those tissues which manifest late radiation damage, shown in Table 3.1). To see this let the individual doses per fraction be denoted by x_{ni}, i = 1,..., n, and let f(x) denote the dose-response relationship appropriate to the tissue under consideration. We assume that f satisfies properties (1), (2), and (3) of Eq. (13), but with strict inequality for (3). We wish to choose the sequence x_{ni}, i = 1,..., n, such that

$$d_n = \sum_{i=1}^{n} x_{ni} - x_1 = \text{maximum}$$

subject to the constraint

$$E = \sum_{i=1}^{n} f(x_{ni}) = \text{constant}$$

The single dose which produces effect E is defined by $x_1 = f^{-1}(E)$. When the undetermined multiplier $c \neq 0$ is introduced, the problem reduces to that of minimizing

$$g(x_{n1},\ldots, x_{nn}) = \sum x_{ni} + c(E - \sum f(x_{ni}))$$

We find

$$\frac{\partial g}{\partial x_{ni}} = 1 - cf'(x_{ni}) = 0, \qquad i = 1,\ldots, n$$

which, by virtue of $f'' < 0$ ((3), Eq. (13)) we may solve to get

$$x_{ni} = x_n \equiv f'^{-1}\left(\frac{1}{c}\right), \qquad i = 1, \ldots, n$$

Finally, the extremum is a maximum, since the minimum $d_n = 0$ is found by choosing one of the x_{ni}, say x_{n1}, equal to x_1, and the rest equal to zero.

Empirically it is found that a lower limit is reached such that further increases in the number of fractions achieve no further sparing (i.e., $\alpha_{eff} \simeq \alpha$). The most relevant survival parameters to radiotherapy are the low-dose response coefficient α (Eqs. (17), (18)) and the dose range over which survival begins decreasing more rapidly due to lethality from accumulated sublethal events [64], that is, passage from the initial, exponential region of the shoulder to the more rapidly decreasing part of the curve. The reasons for this may be understood from Figure 3.6.

We assume that the responses of tissues A and B reflect cell survival, as shown. Moreover, we assume that in a multiple dose schedule, equal response per dose fraction will result (as shown in Figure 3.5). A consequence of the last assumption is that the initial shoulder region will be repeated for each dose fraction, and, given that survival curves A and B are described by linear-quadratic models (Eq. (18)), that the resulting multiple-fraction response curve will be exponential, with slope $-\alpha_{eff} = -\alpha - \beta x$, where x = dose per fraction.

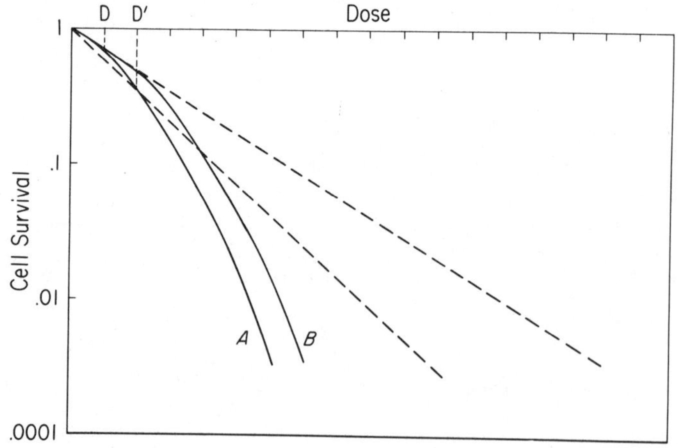

Fig. 3.6. Survival curves with different shoulder widths. Flexure occurs at approximately dose D for tissue A and at D' for tissue B. The maximum difference in response between A and B would be observed following repeated doses equal to D'. The smallest difference would be observed if the dose per fraction were D [from Withers [64], by permission of author and Academic Press, Inc.]

Although there is no dose at which the survival curves begin abruptly to "bend downward," for the practical purposes of radiotherapy it is possible to define the flexure dose, denoted by d_f, as the dose below which the survival curve cannot be statistically resolved from an exponential function of dose. In Figure 3.6, examples of such doses are D for curve A and D' for curve B.

To understand the clinical significance of the dose d_f, suppose curve A were for tumor cells and curve B for dose-limiting normal cells. The maximum favorable differential response would be obtained using doses per fraction D'. This is because the "effective" multifraction survival curves for tissues A and B are exponential, and diverge with respective slopes $-\alpha_A - \beta_A D$ and $-\alpha_B - \beta_B D'$. A decrease in the dose per fraction below D' would result in the same damage to normal tissue for a given total dose, but less damage to the tumor resulting in a smaller therapeutic ratio. Larger dose fractions would result in relatively more normal tissue than tumor damage for the same total dose, again with a decrease in the therapeutic ratios, even if one ignores potentially adverse consequences on tumor response of reducing the number of dose functions.

If curve A (Fig. 3.6) were for cells of a normal tissue and curve B for cells of a tumor, the most *unfavorable* result would be obtained using doses per fraction D'; no favorable differential could be expected at any dose, although the unfavorable result could be minimized using doses per fraction $\leq$ D.

Methods for experimentally determining d_f are described by Withers [64]. One of these, multifraction responses, illustrates an important point in connection with relevant explanatory variables for radiotherapy. The technique is to reduce the size of the dose per fraction in a multiple-fraction schedule that results in fixed effect. By keeping the overall duration of the regimens small, kinetic effects such as proliferation and redistribution (see below) are held at a minimum. It is found that the total dose necessary to achieve this effect increases until the dose per fraction is less than d_f, at which point no further increase in the total dose is necessary when dose per fraction is decreased. The response is then being determined by single-event inactivation ([64], Figure 2).

The important point here is that *the crucial parameter is dose per fraction, not number of fractions* (see discussion of the Ellis "isoeffect" model and Eq. (19) below). Thus, if d_f = 150 rad and the end effect required 1500 rad as 10 fractions each of 150 rad, that effect could be achieved by 1500 rad given in any number of fractions greater than 10 (no increase in total dose). If however, the end effect were a severer injury requiring

3000 rad total dose, then the slope of the multifraction response curve would continue to increase until 20 fractions were delivered.

IN VIVO SURVIVAL DATA (quantitative)

In vivo assays have been developed for three of the principal cell-renewal systems: bone marrow [56], skin [61], and gut [65]. The types of representations discussed in the previous section apply to these data as well, with an important proviso: the number of cells at risk is unknown in the skin and gut assays. Thus the log surviving fraction can only be determined up to an additive constant. In practice, however, the author has not found that this circumstance posed any particular difficulty in the fitting of data. This is especially true in view of the fact that the number of cells at risk is known within an order-of-magnitude (10^4 in the case of the gut assay, 10^6 in the case of the skin assay).

Other types of quantitative *in vivo* methods are those bioassays for the dose which produces a given fixed effect. Thus the estimation of the TCD50 (tumor-control dose for 50% of a population) or LD50 (lethal dose for 50% of a population) is attempted. Note that here the roles of dependent and independent variables have been switched.

IN VIVO RESPONSE DATA (ordinary and binary)

No quantitative measure of cellular damage is possible for this type of data. Responses are ordered, however, and it is possible to test hypotheses using certain statistical methods. An example is the measurement of radiation skin reaction in the mouse foot [11]. Scores of 0, 1, 2, etc. are assigned to no response, slight reddening, dry desquamation, and so on. Berry, Wiernik, and Patterson [8] have used pig skin contraction relative to control fields as a measure of late response (subcutaneous fibrosis). Contraction was measured as the average of four linear dimensions inside a square tatooed on the flank of the animal.

Binary data arise commonly in clinical material, where for example local control (1) and failure (0) are scored for patients categorized according to histopathology, site, age, sex, etc. A dose-response curve can be established for such data using the logistic regression technique of Cox [10]. By comparison of such curves it is possible to establish the relative efficacies of differing therapeutic modalities for tumors at different sites and of different histopathological type.

A clinically well-known model of the Cox logistic-regression type is the so-called NSD model of Ellis [15], where NSD is short for "nominal standard dose." The probability of response, p, is assumed to depend on the explanatory variables N (number of dose fractions, or sessions of radiotherapy), T (overall time of treatment), and D (total given dose) in the following way:

$$\ln\left(\frac{p}{1 - p}\right) = \gamma + \nu \ln N + \tau \ln T + \delta \ln D$$

By fixing p, the following so-called "isoeffect" relationship results:

$$\nu \ln N + \tau \ln T + \delta \ln D = \text{constant}$$

which Ellis wrote in the form

$$\begin{aligned} D &= k(p)\ T^{\alpha}\ N^{\beta} \\ &\equiv \text{NSD} \cdot T^{\alpha} \cdot N^{\beta} \end{aligned} \qquad (19)$$

where NSD = k(p) depends only on p, $\alpha = -\tau/\delta$, and $\beta = -\nu/\delta$. Ellis derived Eq. (19) heuristically, arriving at the estimates $\alpha = .11$, $\beta = .24$ by trial-and-error comparisons with certain clinical data. W. C. J. Hop [29] has derived joint confidence regions for α and β in a generalization of Fieller's theorem.

In recent years the Ellis model (19) has been criticized in connection with both experimental [8, 68] and clinical [2, 6] data. Some of the criticisms summarized by Withers ([63], pp. 267-269) are as follows. First, the formula predicts that the dose D (or the spared dose d_N) for a fixed effect increases without bound as $N \to \infty$, which is true only if the underlying response is described by the multitarget (Eq. (14)) model, which has zero initial slope. If the initial slope is non-zero, as seems to be most often the case, a point is reached when further fractionation does not result in further sparing, and the exponent β for N becomes zero. Second, the time T is relevant to the total dose for a fixed effect mainly as an indicator of the amount of regeneration that can occur during the course of treatment. Some tissues regenerate rapidly, and some very slowly (Table 3.1), and in any event the time of onset and rate of regeneration are dose dependent [20]. Thus a constant value for α cannot hold, even for one tissue. Finally, the version of Eq. (19) purported to apply to tumors has $\alpha = 0$, implying that they do not regenerate, whereas the opposite is known to be true for some animal tumors [3]. The absence or presence of regenerative response is unproved for human tumors.

CYCLE-DEPENDENT RADIATION RESPONSE

Age Response

Cells growing attached to the bottom of Petri dishes become spherical during mitosis, and may be selectively shaken loose and removed [54]. The resulting synchronized population begins moving through the cell cycle, allowing the measurement of age-dependent radiation-survival characteristics. Typical results are shown in Figure 3.7. As seen there, mitosis and G_2 are the most sensitive parts of the cycle, and late S-phase (LS) is the most resistant.

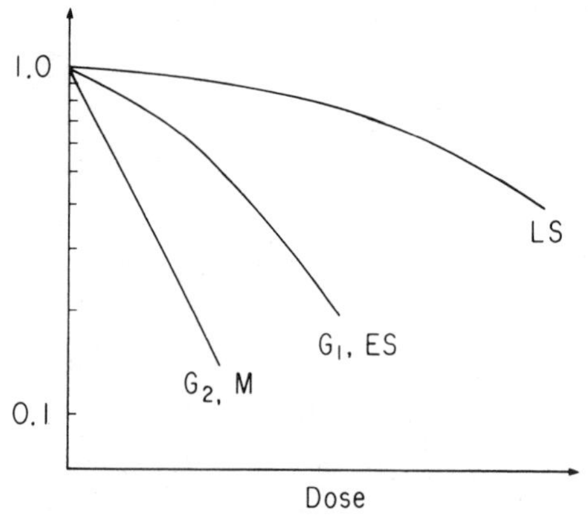

Fig. 3.7. Survival curves for cells in various phases of the cell cycle (ES = early S, LS = late S).

Similar results hold for *in vivo* cell renewal systems. Withers et al. [66] synchronized mouse intestinal stem cells at the G_1/S boundary with hydroxyurea. After release from the block, 1100 rads γ-rays were given at various times. A 100-fold increase in survival occurred between irradiations at 2 hours (early S) and 6 hours (late S).

The corresponding fluctuations in survival parameters have been investigated by Gillespie et al. [18] for the chinese hamster V79 cell line. The results are shown in Figure 3.8 for the linear-quadratic model f_{LQ} (Eq. (18)). α peaks during mitosis and is constant during interphase. β decreases during interphase. The diminished second peaks result presumably from loss of synchrony during progression through the cycle, a phenomenon commonly called "redistribution" (see below).

Because of synchrony decay, the estimation of the age-dependent survival parameters shown in Figure 3.8 is not trivial. Some workers have simply measured survival curves at a fixed sequence of times after selective

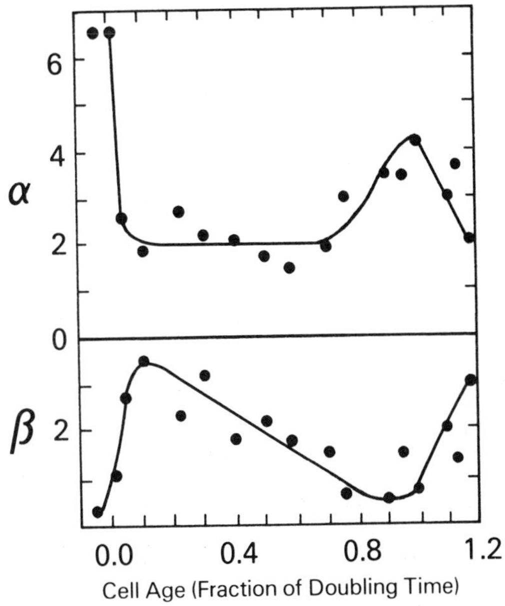

Fig. 3.8. Periodic variation of α and β of the linear-quadratic survival model (18). The doubling time of the population is slightly longer than T_c because of the right skewness of the distribution of cycle transit times. [from Gillespie et al. [18] by permission of author and the Institute of Physics, John Wiley and Sons.]

harvesting of mitotic cells (e.g., [22]), ignoring loss of synchrony. Those who attempt to incorporate its effects run into the difficulty of simultaneous nonlinear estimation of a large number of α's and β's [40]. The following suggested approach retains the desirable features of linear least squares estimation of the dose-response parameters, while allowing correction for desynchronization.

Suppose that at time $t_o = 0$ a pure mitotic population is harvested and allowed to progress through the cycle until the time $t = t_M$ at which a maximum in the number of mitotic figures is observed, corresponding to the passage of the mean of the age density function through M-phase [55]. At times t_i, $i = 0,\ldots, M$, sufficient measurements are taken to construct a survival curve by least squares approximation. Using the Takahashi model Eq. (5), with initial conditions

$$y_j(0) = 0, \qquad j = 1,\ldots, k - 1$$

$$y(0) = 1$$

the population vector $y(t_i)$, $i = 1,\ldots, M$, is computed (the λ's of the coefficient matrix are estimated from PLM data). Survival data at $t = t_0$ yield estimates for α and β characteristic of age k. At $t = t_1$ the cohort will have progressed out of mitosis into G_1. The age density distribution $y_j(t_1)$, $j = 1,\ldots, k$, will be zero for most ages, and cells of these ages will consequently not contribute to survival. To define the ages j where $y_j(t_i) \simeq 0$, $j = 1,\ldots, k$, choose a small positive number ε, say $\varepsilon = 10^{-2}$, and let

$$I_o = 1$$

$$I_i = \min_{k \le j \le k-1} \left\{ j: \sum_{\ell=j}^{k-1} y_\ell(t_i) < \varepsilon \right\}$$

for all $i = 1,\ldots, m$, where

$$m + 1 = \min_{1 \le \mu \le M} \{\mu: y_{k-1}(t_\mu) \le \varepsilon\}$$

Then $\{1, 2,\ldots, k\}$ is the disjoint union of sets J_i, $i = 0,\ldots, m + 1$, where

$$J_i = \{j: I_{i-1} \leq j < I_i\}, \qquad i = 1,\ldots, m$$

$$J_{m+1} = \{j: I_i \leq j < k\}$$

$$J_o = \{k\}$$

We now estimate survival parameters (α_i, β_i) characteristic of each of the age intervals J_i, $i = 0,\ldots, m + 1$. (α_o, β_o) are estimated from survival data for the initially synchronized population:

$$\ell n\ s.f._o\ (x) = -\alpha_o x - \beta_o x^2$$

in which $s.f._o(x)$ denotes surviving fraction at time t_o after dose x. We use these parameters at $t = t_1$ to correct for desynchronization by correcting the observed $s.f._1(x)$. Thus (α_1, β_1) are obtained from survival data at $t = t_1$ by fitting the regression model

$$\ell n[s.f._1(x) - y_k(t_1)\exp(-\alpha_o x - \beta_o x^2)] = -\alpha_1 x - \beta_1 x^2$$

Note that by correcting the observations for desynchronization, we retain the advantages of linearity in the estimation problem for parameters (α_1, β_1) characteristic of ages $j \in J_1$.

Similarly, parameters (α_i, β_i) are estimated by fitting survival data at time t_i to the desynchronization-correlated linear-regression models

$$\ell n[s.f._i(x) - \sum_{\ell=0}^{i-1} \exp(-\alpha_\ell x - \beta_\ell x^2) \sum_{j\in J_\ell} y_j(t_i)] = -\alpha_i x - \beta_i x^2$$

$$i = 1,\ldots, m + 1$$

The consistency of the results may be checked by comparing survival data s.f.(x) from an asynchronous, exponentially growing population with the theoretically predicted values $s.f.(x)_{theoret}$, where

$$s.f.(x)_{theoret} = \sum_{i=0}^{m+1} \exp(-\alpha_i x - \beta_i x^2) \sum_{j\in J_i} y_j(\infty)$$

in which $y(\infty)$ is the asymptotic solution of $dy/dt = Ay$, A given by Eq. (5), with components $y_j(\infty)$ proportional to $2^{-j/k}$.

The periodic sensitivity of a cell population as it progresses through the cycle is usually called its "age-response" function. The parameter variation shown in Figure 3.8 is an example of an age-response. More usually, however, age-responses are presented as survival after a fixed radiation dose as a function of cell age ([45], Fig. 2).

Variation of the parameters β and n of the multitarget model (Eq. (14)) with cell age has been investigated by Fidorra et al. [16], using mouse L cells and pulse cytophotometric analysis to analyze mean phase durations. The extrapolation number n peaked in late S and was ∿constant elsewhere. D_q peaked (Eq. (16)) in early G_1 and late S (resistant parts of the cycle). An important point made by these authors is that the radiosensitivity of G_1 phase seems to depend strongly on its duration. In cells (mouse L) which have a long G_1 phase, G_1 is very radioresistant. The age response functions display a pronounced maximum in the first of G_1 phase. By contrast, chinese hamster cells have a short G_1 that is radiosensitive.

Division Delay (G_2 Block)

Following irradiation of a proliferating population of cells, a period of no growth is observed in which mitotic cells disappear from the population for a time that is proportional to dose (Fig. 3.9). This period has been

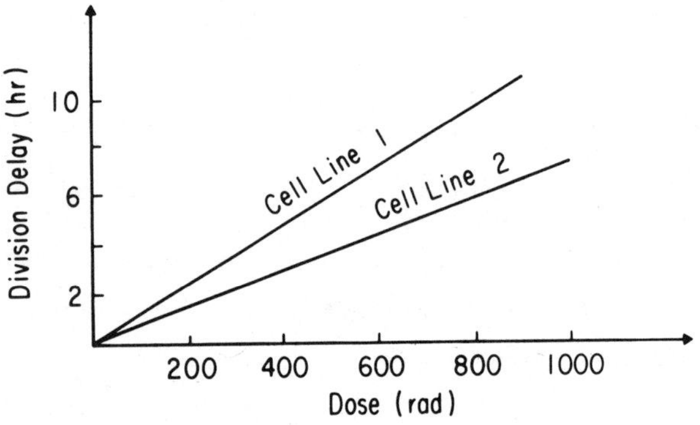

Fig. 3.9. Dose dependence of radiation-induced division delay in two hypothetical cell lines.

termed division delay ([14], pp. 338-380), and seems to result from an accumulation of cells in G_2 (other phenomena may contribute in different cell lines, such as G_1/S block and reduced rate of DNA synthesis in S phase). Both survivors and nonsurvivors undergo the delay (the latter die while attempting division, usually the first or second time following irradiation). Moreover, the delay is age-dependent, with cells in G_2 at the time of irradiation experiencing the longest delays, those in G_1 the least delays, and those in S phase experiencing delays intermediate between these two.

The point at which delay occurs is usually in late G_2 phase (delays have also been measured for G_1), regardless of the cell age at the time radiation is administered [39]. By contrast, Otto and Göhde [41] find a slight prolongation of S-phase accompanying the G_2 block, using pulse-cytophotometric techniques. Interestingly, these authors found qualitative similarity in the division delays effected by x-rays and neutrons.

The large accumulation of cells in G_2 and mitosis does not produce a long-lived synchronization [41]. In the following generation the peak of the G_2 fraction is almost flattened. The authors feel that the rapid decay of synchrony is caused by variation in division delay, not by spread in generation times. Frindel et al. [17] found that 24 hours after 300 rads the age distribution of cells was practically back to normal (as determined using pulse-cytophotometry), while at that time cells were still accumulated in G_1 after 1200 rads. These authors found, in contrast to Mitchell and Bedford [39], that cell kinetics are different after acute irradiation from those observed after or during low-dose-rate continuous irradiation.

Redistribution

The variability in the age-response shown in Figures 3.7 and 3.8 indicates that the survival of cells in an asynchronous population (i.e., one having a stable exponential age distribution) exposed to clinically important doses of x-irradiation (200-300 rads) will vary greatly, depending on their distribution within the cycle. Another way of saying this is that a perturbed age-distribution results, which is resistant relative to the asynchronous population. If the next dose (in a multifraction series) is delayed, surviving cells progress through the cycle. Their response will fluctuate with the time interval.

The rate at which cells progress through the cycle varies, not only from tissue to tissue but also within the same tissue (i.e., T_c has a probability distribution in a homogeneous cell population). Thus, there is a *progressive loss of synchrony*, which is called *redistribution*. It may be clearly pictured in terms of the Hahn model (Eqs. (8), (9)).

The age distribution following irradiation at time t = 0 is given by [22],

$$y(0) = R(x)y(\infty) \tag{20}$$

in which $y(\infty)$ is the stable age distribution and R(x) is a diagonal matrix whose (i, i) element is the survival response of the population at age i to radiation dose x. For example, using the age dependence of the linear-quadratic parameters α and β shown in Figure 3.8,

$$R(x)_{ii} = \exp(-\alpha_i x - \beta_i x^2), \qquad 1 \leq i \leq k \tag{21}$$

in which the division cycle has been divided into k compartments (Eqs. (9), (10)).

Division delay may be described by defining a G_2M-block "operator" C, which is similar to matrix B of Eq. (9) except that the last few diagonal entries (corresponding to the G_2M ages) are ones, with zeros off-diagonal. If we define τ as the coefficient of division delay in Figure 3.9, we have

$$\text{division delay} = \frac{\tau x}{\Delta} = j \text{ time units}$$

where Δ is the time step chosen for the Hahn model. Assuming for simplicity that j = integer, we have for the age distribution at time $t = n\Delta$:

$$y(n) = \begin{cases} C^n\, y(0), & 0 \leq n \leq j \\ B^{n-j}\, C^j y(0), & n > j \end{cases} \tag{22}$$

Equations (20)-(22) define the sequence of states y(0), y(1),... which describe the return of the radiation perturbed age distribution toward a multiple of $y(\infty)$.

If desynchronization between radiotherapeutic dose fractions returns surviving cells in a tissue to their pre-radiation cell-cycle distribution, the tissue will be sensitized relative to its state if there had been no redistribution. In terms of the quantities defined in Eqs. (20)-(22), this means that the surviving fraction of the state y(0) after a dose x will be greater than that of the state y(∞):

$$\frac{\sum_{i=1}^{k} \left[R(x)\ y(0)\right]_i}{\sum_{i=1}^{k} y(0)_i} > \frac{\sum_{i=1}^{k} \left[R(x)\ y(\infty)\right]_i}{\sum_{i=1}^{k} y(\infty)_i}$$

Since modern radiotherapy using protracted treatment times is limited by the tolerance of such relatively nonproliferative (and therefore nonredistributing) tissues as the fibrovascular connective tissues, cartilage, bone, or spinal cord [62], redistribution may be an important factor determining the therapeutic advantage of dose fractionation. "Nonproliferative" is taken to mean that generation times are long compared with duration of therapy. (See Table 3.1.) Tumors may be expected to undergo redistribution toward radiosensitive states more rapidly than the dose-limiting, nonproliferative normal tissues listed in Table 1.

Regeneration

After sustaining radiation injury, cell renewal systems respond to the resulting depopulation by increasing their rate of proliferation, i.e., by changing the balance between cell loss and cell production (see Eq. (12)) in favor of the latter. Both normal tissues and tumors show this response.

The ability of tissue and tumor to regenerate after radiation injury depends on survival of cells capable of proliferation. Tissues such as the central nervous system, cartilage, bone, and connective tissues show little or no regenerative response to radiation, whereas bone marrow, gut, oropharyngeal mucosae, and skin, which normally proliferate actively maintaining cell production and loss in balance, will manifest depopulation and compensatory repopulation during the course of radiotherapy. Injury sustained by the former (nonproliferating) group is usually observed long after the end of therapy, and is thus called *late response*. Late responses are usually dose-limiting in external beam radiotherapy at energies greater than 1 MeV (which for technical reasons are relatively sparing of the skin). Injury sustained by the cell renewal group, which also includes most tumors, is manifested early on and is thus called *acute response*.

When radiation dose is fractionated into multiple small daily doses, many normal cell renewal systems are able to maintain themselves against the repeated insult by means of compensatory repopulation. This is illustrated for mouse gut by the results shown in Figure 3.10 [20]. The crucial effects of dose and time of delivery are evident. Panel E illustrates the usual pattern of exposure used clinically. Proliferative activity varies between 40% of control to somewhat above control (because of the Friday-Monday radiation-free interval). For the entire week, however, integrated proliferative activity was 83% of control. Panel F illustrates what happens when the compensatory response is retarded by giving radiation at the time it begins to occur. As indicated by the integrated proliferative activity (69%), overall regeneration was markedly compromised. The other panels (A-D) are illustrative of the burst of proliferative activity that follows radiation given early in the week.

As the course of therapy extended into the fourth week, the picture changed (Figure 3.11), with heightened p (proliferative) cell activity. Panel A shows that conventional daily fractionation did not result in net depressed proliferation, probably due to "recognition" of damage incurred during previous weeks. Again, exposures timed to coincide with proliferative bursts resulted in substantial reduction in net cell production (Panel D).

The bursts of proliferative activity shown in Figures 3.10 and 3.11 result from two factors: (1) retention in cycle of cells otherwise committed to maturation and (2) shortening of cell-cycle time T_c of the p cells. Similar factors operate in the proliferative response of tumors to fractionated irradiation [3], although there appears to be considerable variation between tumors ([26], p. 164).

The response of tumors to fractionated irradiation may be slower than that of normal cell renewal systems, as is illustrated by Figure 3.12 [3]. Curve e (lower panel) shows the variation of the fraction of clonogenic cells in the tumors as a function of time after the start of five times weekly treatment with 200 rad/fraction 300 kV x-rays. Note that the proliferative cells of the tumor are unable to maintain their baseline value

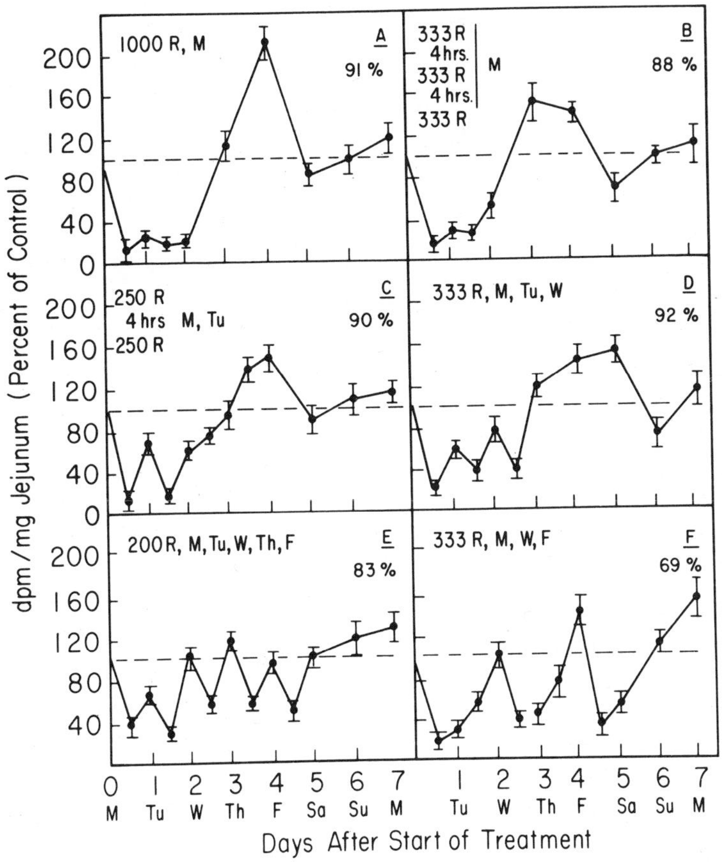

Fig. 3.10. Intestinal cell proliferation during fractionated abdominal irradiation. Ordinates show relative proliferative activity as a function of time after start of treatment on Monday (M). Exposure schedules are shown at upper left of each panel, and percentages at right indicate integrated proliferative activity in percent of control values (thus in the lower left-hand panel, 200 rad were given Monday through Friday, resulting in a decrease of proliferative activity to 83% of control value) [from Hagemann [20], with permission of author and the British Journal of Radiology].

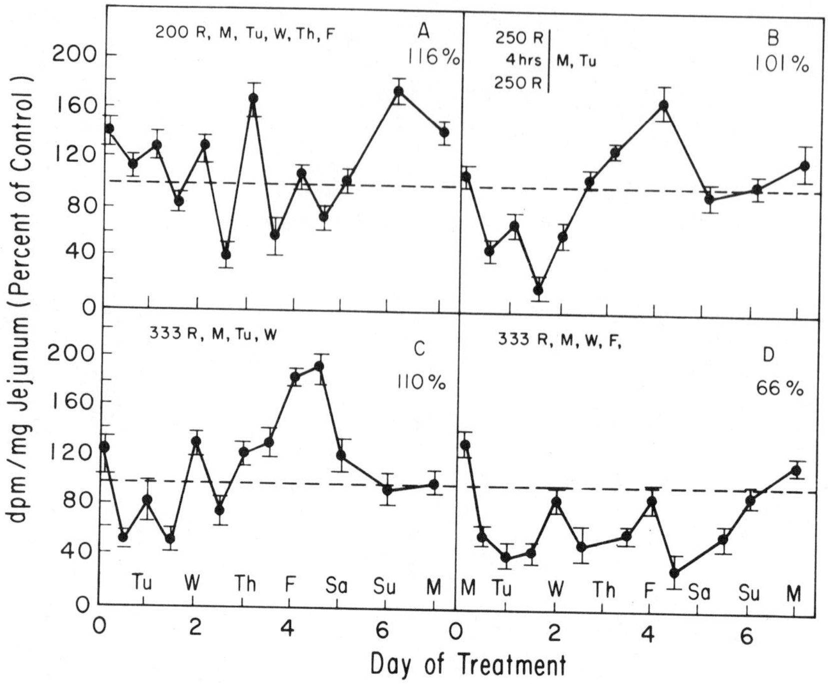

Fig. 3.11. Same as Figure 3.10, except the one-week measurement period came after three weeks of the indicated treatment schedules [from Hageman [20], with permission of author and the British Journal of Radiology].

against daily injury by 200 rad x-rays, as opposed to the situation for gut (Figures 3.10 and 3.11; although the ordinates are not strictly comparable, we interpret "dpm/mg jejunum" to be indicative of the number of surviving p-cells). That the p cells of the tumor have entered an active phase of proliferative response is indicated by the rapid upward movement of the curve after cessation of therapy (open circles).

Other important points emerge from the data shown in Figure 3.12:

1. Regeneration of clonogenic cells (those capable of an unlimited number of divisions) is not reflected by volume changes (curve b) in the irradiation tumor; in fact, growth continued 7 days after treatment started, a period during which clonogenic cells were sharply diminished in number, and a week's delay in volume increase followed the upsurge in clonogenic cell number after treatment was stopped.

2. During the first week of treatment the fraction of p cells decreased more rapidly than in the second week, and in the third week the fraction remained roughly constant. The same diminution in effectiveness of treatment in the second and third weeks was observed with neutrons. Barendsen and Broerse [3] argue that most of the decrease in effectiveness must be due to increased rate of proliferation of clonogenic tumor cells.

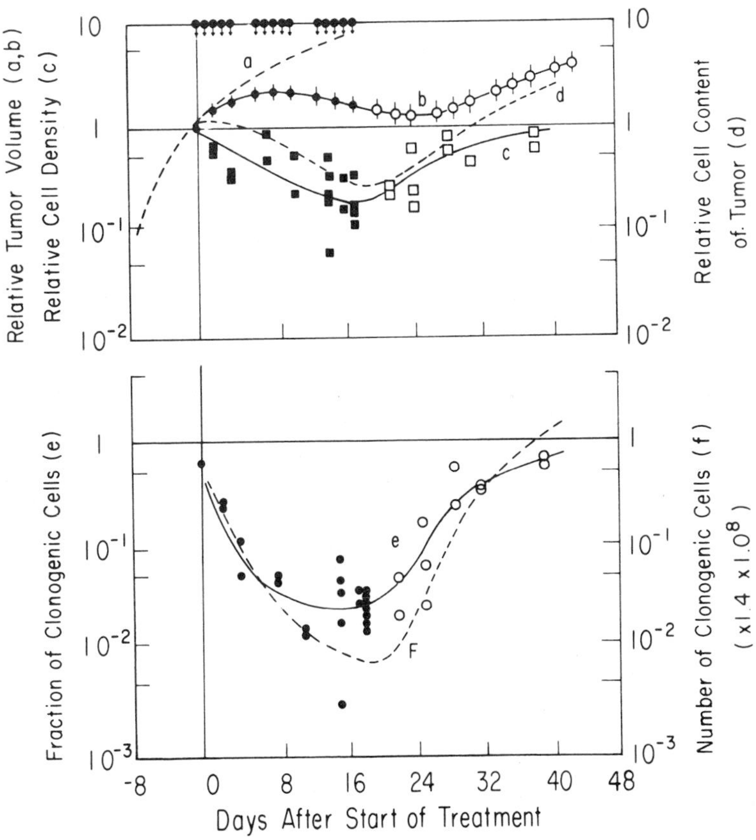

Fig. 3.12. Changes in tumor volume, number of cells, and numbers of clonogenic cells in experimental rhabdomyosarcomas treated five times weekly for three weeks with 200 rad per fraction of 300 kV X-rays. Closed (open) symbols refer to data obtained during (after) treatment. Mean volume at start of treatment 0.8cc, mean cell content 1.4×10^8 cells.

Curve a: Growth of controls, relative to initial volume.

Curve b: Growth of irradiated tumors, relative to initial volume. Arrows indicate times of irradiation.

Curve c: Number of intact cells per gram of tumor, relative to control of same volume.

Curve d: Number of intact cells in tumor, relative to number at start of treatment.

Curve e: Fraction clonogenic cells relative to control.

Curve f: Number of clonogenic cells in tumor.

[from Barendsen and Broerse [3], with permission of authors and Pergamon Press, Inc.].

These results may be incorporated into Eq. (22) by using the assumption that flow rates respond to the total population count, as Jansson [31] did (Eq. (11a)). Thus, each of the off-diagonal elements β and γ of the matrix B (Eq. (9)) would be modified in the following way for time $t = n\Delta$:

$$\begin{aligned} \beta(n) &= \beta u(\xi) \\ \gamma(n) &= \gamma u(\xi) \\ Q(n) &= Qv(\xi) \end{aligned} \tag{23}$$

In Eq. (23) u and v are decreasing functions of the variable

$$\xi = \lambda^{-n} \sum_{\ell=1}^{k} y(n)_{\ell}$$

and λ is the largest eigenvalue of the steady-state matrix ($\xi = 1$), given by Eq. (9). The function $u(\xi)$ has the properties

$$\begin{aligned} &u(1) = 1 \quad \text{(steady-state growth)} \\ &u(\xi) \leq (\beta + \gamma)^{-1} \\ &u'(\xi) < 0 \end{aligned} \tag{24}$$

in which β and γ are the steady-state values of the transition probabilities of the Hahn model, (Eq. (8)). Some choices for u (selection would be made based on fit to data) are shown in Figure 3.13. Immediately after irradiation, ξ is the surviving fraction of clonogenic cells. After mitotic delay, the rate of proliferation increases. The lower bound on the transit time of the mean of the age p.d.f., T_M (Eq. (11)), is

$$T_M(n) \geq (\beta + \gamma)T_M \quad \text{(steady state)}$$

The function $v(\xi)$ satisfies

$$\begin{aligned} &v(1) = 1 \quad \text{(steady-state growth)} \\ &v(\xi) \leq 2/Q \\ &v'(\xi) < 0, \end{aligned} \tag{25}$$

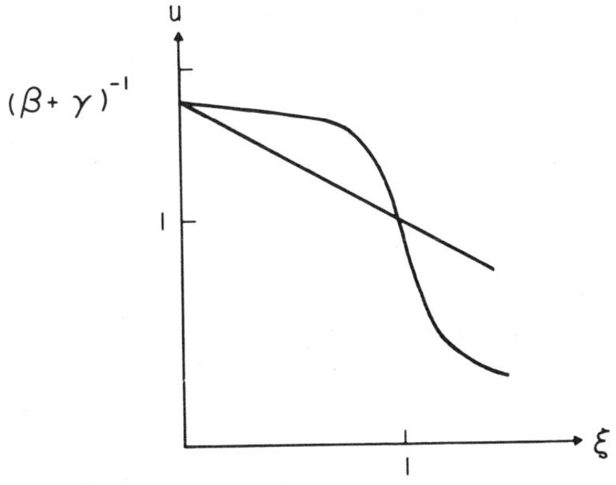

Fig. 3.13. Two possibilities for $u(\xi)$ (u alters the flow rate through the cell cycle in response to a radiation-induced decrease in the number of proliferative cells; text Eqs. (23)-(24)).

where $Q - 1 \in (0, 1)$ is the growth fraction characteristic of steady-state conditions.

By simulating the growth curve of an irradiated population using the altered Hahn model (Eqs. (8), (9), (20)-(22), (23)-(25)), the slopes m_u and m_v of linear u and v may be determined. Linear forms may be rejected if, for example, $m_u < 1 - (\beta + \gamma)^{-1}$ and $m_v < 1 - 2/Q$. An independent check is to fit PLM data of the irradiated tissue [36] with parameters determined from growth curves.

MODELS OF MULTIFRACTION RADIOTHERAPY

In this concluding section we review some of the models that have been proposed for the coupling between cell-cycle related phenomena and radiation injury. Since mathematical representations of the cell cycle and radiation dose responses have been presented, the description will be cast in terms of them. Also in the interest of brevity, only a limited selection of work published in this area over the last decade will be mentioned.

Broadly speaking, the modeling has had as its goal either faithful simulation of cell-cycle dependent radiation response or control theory approaches to the optimization of treatment, assuming "complete" knowledge of tissue response to radiation. In the former category, Hahn and Kallman [22] successfully simulated the response of mammalian cells in culture to single and double doses of X-rays. The effects of division delay and differential age sensitivity were taken into account, using formulations similar to those found in Eqs. (8)-(20), (14), and (20)-(22). Steward and Hahn [50] used the same methods in applying information concerning age-response functions and mitotic delay to the "optimization" (no methods of control theory were employed) of treatment schedules for both radiation and drugs. Significantly, the effect of transferring cells between cycling and noncycling compartments was included in the treatment of chemotherapy. Also, the therapeutic ratio (tumor to normal cell kill) was computed for cycling normal cells in the case of X-ray therapy. No account of differences in early and late normal tissue responses have been included in these or other studies known to the author.

Hethcote and Waltman [27] and Almquist and Banks [1] apply optimal control theory to the problem of fractionated radiotherapy of tumors which are heterogeneous in regard to the oxygenation status of their cells (thus, cell-cycle dependent effects are ignored). The rationale for these treatments lies in the higher sensitivity of well-oxygenated cells to radiation damage. Moreover, it is known that the fraction of oxic cells changes as a result of treatment, with previously hypoxic cells replacing "dead" oxic ones by becoming oxygenated. This phenomenon has been termed *reoxygenation*. Elkind [12] has put forth an optimization strategy based on the kinetics of reoxygenation. He points out that the sparing effects of dose fractionation (Fig. 3.5) can be cancelled by the effects of reoxygenation. This represents an important potential differential, since normal cells are well-oxygenated. It requires, however, that the effects of tumor repopulation be ignored.

The primary goal of the papers of both Hethcote and Waltman (henceforth H-W) and Almquist and Banks (henceforth A-B) was to devise an optimal control theory framework within which *any model* of the dose response of irradiated tissues could be used to make predictions about optimal treatment plans. Since this goal has now largely been realized, it is natural to turn next to the problem of devising better models of radiation response. In order to illustrate some of the most promising directions for development as seen by this author, the models used by H-W and A-B will be presented in some detail, with discussion and suggestions for improvement. The presentation will be an assimilation of the models used by H-W and A-B, and will differ in some respects from each.

The state variables used are

N_1 = live, oxygenated cell count,
N_2 = dead, oxygenated cell count,
N_3 = live, anoxic cell count,
N_4 = dead, anoxic cell count, and
N_5 = tumor-bed (normal tissue) cell count.

The single-hit, multitarget model (14) is used to describe cell killing by radiation, with various parameters assumed for the responses of the five types of cells. The growth (A-B) after radiation dose x at time t_o is assumed to be represented by

$$N_1(t - t_o) = \begin{cases} N_1(t_o), & t < \gamma x \\ N_1(t_o) \exp[\alpha(t - \gamma x)], & t \geq \gamma x \end{cases} \tag{26}$$

with $-\alpha$ replacing α in the equation for N_2. Anoxic cells do not proliferate. Whereas (A-B) ignore proliferation of normal cells, (H-W) assume that after one day the normal cells increase as described by the "logistic" function:

$$N_5(t - t_o) = \begin{cases} \dfrac{H\ N_5(t_o)}{N_5(t_o) + [H - N_5(t_o)] \exp[-H\delta(t - 1)]}, & t \geq t_o + 1 \\ N_5(t_o), & t < t_o + 1 \end{cases} \tag{27}$$

in which t_o is the time of treatment and δ and H are parameters. The logistic function is a solution of a nonlinear Ricatti equation:

$$\frac{dN}{dt} = \delta\ N(H - N)$$

The cost functionals used were number of cells remaining in the tumor bed (H-W) and a linear combination of N_1, N_3, $[\ell n(N_5)]^2$, and penalties and costs associated with treatment (A-B). The class of acceptable controls (doses per fraction, between-treatment times, and number of fractions) were restricted in each case. The problem was to drive the number of viable tumor cells to less than 1 (H-W) while maximizing the number of cells in the tumor bed (normal tissue). (A-B) assumed a fixed treatment "calendar," and determined the set of doses per fraction which extremized their cost functional.

The omission of age-specific radiosensitivities may not be a serious shortcoming in models of this type. The parameters will never be known for all the normal cells of interest, much less for the tumors, for which heterogeneity is the rule, both between tumors of the same size and histological type and within a given tumor [26]. If, however, the status of a slowly proliferating normal tissue were of interest (as is always the case in radiotherapy), its lack of *redistribution* toward more sensitive states following radiation, and the tendency of both tumor and regenerating normal cells to tend toward such states, would both be factors to reckon with in considering the therapeutic efficiency of fractionated radiotherapy [62].

More serious difficulties, in this author's view, are the following. First, the multitarget model (14) is almost certainly not correct for either tumor or normal tissue, as it violates the first of the conditions put forth in Eq. (13). Referring to Fig. 3.5 it may be seen that the effective (negative) slope, α_{eff}, of the multifraction response curve has no upper limit for the multitarget model, since $f'_{MT}(0; \beta, n) = 0$. Thus the dose for a given effect will increase without limit as the number of daily treatments increases, in contrast to the empirical finding that an effective lower dose limit, d_f ("flexure dose"), exists such that further increases in total dose to achieve a given effect are necessary after the dose per fraction decreases below d_f. The Ellis model (19) above is consistent with the unlimited increase in total dose for a given effect. From Table 1 of (A-B) the following approximately equal-effect [total dose, number of fractions] pairs may be extracted from the "Optimal equal-dose program" grouping: [5727, 25], [6301, 30], [6836, 35], [7355, 40]. Referring to Eq. (19), we plot these points in log-log coordinates, assuming overall time proportional to number of treatments. They lie perfectly on a line with slope 0.53, somewhat higher than Ellis' 0.35 [15].

Second, the equations (26) for tumor growth and (27) for normal tissue *regenerative response* to radiation do not tell the story correctly. To start with, the normal tissues of most concern, those which manifest so-called "late injury" months or years following treatment (Table 3.1), do not proliferate at all, or if they do so, to a negligible extent. Moreover, normal renewal system response is not logistic, as in Eq. (27), but rather overshoots the steady-state population level to a large extent [36], and the level of this dose-dependent overshoot is critical to the ability of regenerating normal cells to maintain themselves against repeated radiation insult [20]. There are dose-dependent delays (5-20 hours after G_2 delay) before any proliferative response begins. Tumors show the same characteristics, in a qualitative sense, as normal cell renewal systems [3], but may be unable to cope with the same doses of repeated radiation (cf. Fig. 3.10-3.12). Finally, the use of models like Eqs. (26) and (27) is inconsistent with the delicate balance between *cell production* and *cell loss* that describe both normal and malignant renewal systems [47].

Third, the dose-limiting late response is not included in the cost functional. It is not obvious how this may be done, given the arbitrary nature of dose-response scales which characterize assays for long-term injury (see e.g., [2, 8, 68]). It is known that some repair of sublethal injury takes place (i.e., $f''(x) < 0$ in Eq. (13)), so that decreasing the dose per fraction leads to some sparing of tissues that will manifest late damage. However, quantitative estimates of parameters for the appropriate dose-response curve are not available.

Finally, the pronounced heterogeneity of tumors, even when of the same size and of the same histological type, argues against attempting to model in any detail their response to fractionated radiation. Indeed, the very processes that are selected for modeling are often, as in the case of the oxygen effect, of questionable importance in the treatment of human tumors [66]. More promising in the author's view is an approach that concentrates on normal tissues, both proliferative and nonproliferative, in an attempt to find fractionation schedules which maximize tumor dose while satisfying given constraints on early and late damage. There is a conflict in the requirements for increased fractionation (decreased late response) and longer periods between dose fractions (increased proliferative recovery of cell-renewal systems). The modeling required has yet to be attempted.

ACKNOWLEDGMENTS

This article is extracted from the presentation "Cell Cycles and Radiation Damage" (National Science Foundation Conference Board on Mathematical Sciences, Carbonale, Ill., June, 1978, H. T. Banks, principal lecturer).

I am greatly indebted to Dr. S. L. Rasmussen and Dr. S. O. Zimmerman (Biomathematics), to Dr. L. J. Peters and Dr. H. R. Withers (Experimental Radiotherapy), and to Dr. G. H. Fletcher (Radiotherapy) for the careful attention they devoted to this manuscript, resulting in many corrected errors and improvements in presentation. The contribution of Dr. Withers' ideas concerning multifraction radiotherapy are obvious throughout. For permission to reproduce figures, I owe thanks to Dr. H. R. Withers and Academic Press, Inc., to Dr. C. J. Gillespie and The Institute of Physics, John Wiley and Sons, and to Dr. R. F. Hagemann and The British Journal of Radiology. The support of the facilities of the Department of Biomathematics, M. D. Anderson Hospital and Tumor Institute was indispensable in the preparation of this manuscript, as was the patience of Ms. Betty Schwarz during the many revisions that were necessary, throughout which she maintained the same high standards.

This work was funded by Grant CA-11430 (Biomathematics in a Cancer Institute) from the National Cancer Institute.

REFERENCES

1. K. J. Almquist and H. T. Banks, "A theoretical and computational method for determining optimum treatment schedules in fractionated radiotherapy," *Math. Biosci.* 29(1976), 159-179.

2. G. Arcangeli, M. Friedman and R. Paoluzi, "A quantitative study of late radiation effect on normal skin and subcutaneous tissues in human beings," *Br. J. Radiol.* 47(1974), 44-50.

3. G. W. Barendsen and J. J. Broerse, "Experimental radiotherapy of a rat rhabdomyosarcoma with 15 MeV neutrons and 300 kV X-rays. II. Effects of fractionated treatments, applied five times a week for several weeks," *Eur. J. Cancer* 6(1970), 89-109.

4. J. C. Barrett, "A mathematical model of the mitotic cell and its application to the interpretation of percentage labeled mitoses data," *J. Natl. Cancer Inst.* 37(1966), 443-450.

5. M. S. Bartlett, "Distributions associated with cell populations," *Biometrika* 56(1969), 391-400.

6. T. D. Bates and L. J. Peters, "Danger of the clinical use of the NSD formula for small fraction numbers," *Br. J. Radiol.* 48(1975), 773.

7. H.-P. Beck, "A new analytical method for determining duration of phases, rate of DNA synthesis and degree of synchronization from flow-cytometric data on synchronized cell populations," *Cell Tissue Kinet.* 11(1978), 139-148.

8. R. J. Berry, G. Wiernik and T. J. S. Patterson, "Skin tolerance to fractionated X-irradiation in the pig--how good a predictor is the NSD formula?" *Br. J. Radiol.* 47(1974), 185-190.

9. M. Collyn-d'Hooghe, A.-j. Valleron and E. P. Malaise, "Time-lapse cinematography studies of cell cycle and mitosis duration," *Exp. Cell Res.* 106, 2(1977), 405-407.

10. D. R. Cox, *The Analysis of Binary Data*, Methuen and Co. Ltd., London, England (1970).

11. J. Denekamp and J. F. Fowler, "Further investigations of the response of irradiated mouse skin," *Int. J. Radiat. Biol.* 10(1966), 435-441.

12. M. M. Elkind, "Recovery, reoxygenation and a strategy to improve radiotherapy," in *The Biological and Clinical Basis of Radiosensitivity*, M. Friedman (Ed.), Charles C. Thomas, Springfield, Ill. (1974), 343-372.

13. M. M. Elkind, "The initial part of the survival curve," *Radiat. Res.* 71(1977), 9-23.

14. M. M. Elkind and G. F. Whitmore, *The radiobiology of cultured mammalian cells*. Gordon and Breach, New York (1967), 1-51.

15. F. Ellis, "Nominal standard dose and the ret," *Br. J. Radiol.* 44(1971), 101-108.

16. J. Fidorra and W. A. Linden, "Radiosensitivity and recovery of mouse L cells during the cell cycle," *Rad. and Environm. Biophys.* 14(1977), 285-294.

17. E. Frindel, G. M. Hahn, D. Robaglia and M. Tubiana, "Responses of bone marrow and tumor cells to acute and protracted irradiation," *Cancer Res.* 32(1972), 2096-2103.

18. C. J. Gillespie, J. D. Chapman, A. P. Reuvers and D. L. Dugle, in *Cell survival after low doses of radiation: theoretical and clinical implications*, T. Alper (Ed.), Institute of Physics and John Wiley, London, England (1975), 25-33.

19. J. W. Gray, "Cell-cycle analysis of perturbed cell populations: computer simulation of sequential DNA distributions," *Cell Tissue Kinet.* 9(1976), 499.

20. R. F. Hagemann, "Intestinal cell proliferation during fractionated abdominal irradiation," *Br. J. Radiol.* 49(1976), 56-61.

21. G. M. Hahn, "State vector description of the proliferation of mammalian cells in tissue culture. I. Exponential growth," *Biophys. J.* 6(1966), 275-290.

22. G. M. Hahn and R. F. Kallman, "State vector description of the proliferation of mammalian cells in tissue culture," *Radiat. Res.* 30(1967), 702-713.

23. T. E. Harris, "A mathematical model for multiplication by binary fission," in *The Kinetics of Cellular Proliferation*, F. Stohlman, Jr. (Ed.), Grune and Stratton, New York (1959), 368-381.

24. T. E. Harris, *The Theory of Branching Processes.* Springer-Verlag, Berlin (1963).

25. N. R. Hartman and T. Pedersen, "Analysis of the kinetics of granulosa cell populations in the mouse ovary," *Cell Tissue Kinet.* 3(1970), 1.

26. A. F. Hermens, *Variations in the cell kinetics and the growth rate in an experimental tumor during natural growth and after irradiation.* (Ph.D. thesis, University of Amsterdam). Published by the Radiobiological Institute of the Organization for Health Research TNO, Rijswijk, Netherlands (1970), 76-98.

27. H. W. Hethcote and P. Waltman, "Theoretical determination of optimal treatment schedules for radiation therapy, *Radiat. Res.* 56(1973), 150-161.

28. H. W. Hethcote, J. W. McLarty and H. D. Thames, Jr., "Comparison of mathematical models for radiation fractionation," *Radiat. Res.* 67(1976), 387-407.

29. W. C. J. Hop, Personal Communication (1978).

30. A. Howard and S. R. Pelc, "Synthesis of deoxyribonucleic acid in normal and irradiated cells and its relation to chromosome breakage," *Heredity* (*Suppl.*) 6(1952), 261.

31. B. Jansson, "Competition within and between cell populations," in *Fraction Size in Radiobiology and Radiotherapy*, T. Sugahara, L. Révész and O. Scott (Eds.), Igaku Shoin Ltd., Tokyo (1973), 51-72.

32. B. Jansson, "Simulation of cell-cycle kinetics based on a multicompartment model," *Simulation* 25(1975), 99-108.

33. A. M. Kellerer and H. H. Rossi, "The theory of dual radiation action," *Current Topics Radiat. Res. Q.* 8(1972), 85-158.

34. D. G. Kendall, "On the role of variable generation time in the development of a stochastic birth process," *Biometrika* 35(1948), 316-330.

35. B. Klein and A.-j. Valleron, "A compartmental model for the study of diurnal rhythms in cell proliferation," *J. Theoret. Biol.* 64(1977), 27-42.

36. S. Lesher, J. Cooper, R. Hagemann and J. Lesher, "Proliferative patterns in the mouse jejunal epithelium after fractionated abdominal X-irradiation," *Current Topics Radiat. Res. Q.* 10(1975), 229-261.

37. P. D. M. MacDonald, "Statistical inference from fraction labelled mitoses curves," *Biometrika* 57(1970), 489-503.

38. P. D. M. MacDonald, "On the statistics of cell proliferation," in *The Mathematical Theory of the Dynamics of Biological Populations*, M. S. Bartlett and R. Hiorns (Eds.), Academic Press, London (1973).

39. J. B. Mitchell and J. S. Bedford, "Dose-rate effects in synchronous mammalian cells in culture. II. A comparison of the life cycle of

HeLa cells during continuous irradiation or multiple-dose fractionation," *Radiat. Res.* 71(1977), 547-560.

40. J. Niederer and J. R. Cunningham, "The response of cells in culture to fractionated radiation: a theoretical approach," *Phys. Med. Biol.* 21(1976), 823-839.

41. F. Otto and W. Göhde, "Effects of fast neutrons and X-ray irradiation on cell kinetics," in *Pulse-Cytophotometry*, W. Göhde, J. Schumann and T. Büchner (Eds.), European Press, Ghent, Belgium (1976), 244-249.

42. H. Quastler and F. G. Sherman, "Cell population kinetics in the intestinal epithelium of the mouse," *Exp. Cell Res.* 17(1959), 420-438.

43. J. L. Roti Roti and L. A. Dethlefsen, "Matrix simulation of duodenal crypt cell kinetics. II. Cell kinetics following hydroxyurea," *Cell Tissue Kinet.* 8(1975), 335-353.

44. S. I. Rubinow, "Maturity-time representation for cell populations," *Biophys. J.* 8(1968), 1055-1073.

45. W. K. Sinclair, in *Conference on Time and Dose Relationships in Radiation Biology as Applied to Radiotherapy*. Springfield, Va. Clearinghouse Fed. Sci. Tech. Inf., Brookhaven Nat. Lab. Rep. BNL 50230 (C-57) (1970), 97-116.

46. G. G. Steel, "Cell loss as a factor in the growth rate of human tumors," *Eur. J. Cancer* 3(1967), 381-387.

47. G. G. Steel, "Cell loss from experimental tumors," *Cell Tissue Kinet.* 1(1968), 193-207.

48. G. G. Steel and L. F. Lamerton, in *Human Tumor Cell Kinetics*, (National Cancer Institute Monograph 30) S. Perry (Ed.), US Government Printing Office, Washington, D. C. (1969), 42.

49. G. G. Steel, K. Adams and J. C. Barrett, "Analysis of the cell population kinetics of transplanted tumors of widely-differing growth rate," *Br. J. Cancer* 20(1966), 784-800.

50. P. G. Steward and G. M. Hahn, "The application of age-response functions to the optimization of treatment schedules," *Cell Tissue Kinet.* 4(1971), 279-291.

51. R. N. Stuart and T. C. Merkle, *The calculation of treatment schedules for cancer chemotherapy - Part II*. University of California LRL Report UCRL-14505. University of California, Berkeley, Ca. (1965).

52. D. E. Swartzendruber, "A BUdR-mithramycin technique for detecting cycling and noncycling cells by flow microfluorometry," *Exp. Cell Res.* 109 (1977), 439.

53. M. Takahashi, "Theoretical basis for cell cycle analysis. I. Labelled mitosis wave method," *J. Theoret. Biol.* 13(1966), 202-211.

54. T. Terasima and L. J. Tolmach, "Variations in several responses of HeLa cells to X-irradiation during the division cycle," *Biophys. J.* 3(1963), 11-33.

55. H. D. Thames, Jr. and R. A. White, "State-vector models of the cell cycle. I. Parametrization and fits to labelled mitoses data," *J. Theoret. Biol.* 67(1977), 733-756.

56. J. E. Till and E. A. McCulloch, "A direct measurement of the radiation sensitivity of normal mouse bone marrow cells," *Radiat. Res.* 14(1961), 213-222.

57. E. Trucco and P. J. Brockwell, "Percentage labelled mitoses curves in exponentially growing cell populations," *J. Theoret. Biol.* 20(1968), 321-337.

58. H. von Foerster, in *The Kinetics of Cellular Proliferation*," F. Stohlman, Jr. (Ed.), Grune and Stratton, New York (1959), 382.

59. R. A. Walters and D. F. Pedersen, "Radiosensitivity of mammalian cells. I. Timing and dose-dependence of radiation-induced division delay," *Biophys. J.* 8(1968), 1475-1486.

60. G. H. Weiss, "Equations for the age structure of growing populations," *Bull. Math. Biophys.* 30(1968), 427-435.

61. H. R. Withers, "The dose-survival relationship for irradiation of epithelial cells of mouse skin," *Br. J. Radiol.* 40(1967), 187.

62. H. R. Withers, "Cell-cycle redistribution as a factor in multifraction irradiation," *Radiology* 114(1975a), 199-202.

63. H. R. Withers, "The four R's of radiotherapy," *Advances Radiation Biology* 5(1975b), 241-271.

64. H. R. Withers, "Tissue responses to multiple small dose fractions," *Radiat. Res.* 71(1977), 24-33.

65. H. R. Withers and M. M. Elkind, "Radiosensitivity and fractionation response of crypt cells of mouse jejunum," *Radiat. Res.* 38(1969), 598-613.

66. H. R. Withers and H. D. Suit, "Is oxygen important in the radio-curability of human tumors?" in *The Biological and Clinical Basis of Radiosensitivity*, M. Friedman (Ed.), Charles C. Thomas, Springfield, Ill. (1974), 548-560.

67. H. R. Withers, A. M. Chu, B. O. Reid and D. H. Hussey, "Response of mouse jejunum to multifraction irradiation," *Int. J. Radiat. Onc., Biol., Phys.* 1(1975), 41-52.

68. H. R. Withers, H. D. Thames, Jr., B. L. Flow, K. A. Mason and D. H. Hussey, "The relationship of acute to late skin injury in 2 and 5 fraction/week γ-ray therapy," *Int. J. Radiat. Onc., Biol., Phys.* 4 (1978), 595-602.

69. H. R. Withers, K. Mason, B. O. Reid, N. Dubravsky, E. T. Barkley, B. W. Brown and J. B. Smathers," Response of mouse intestine to neutrons and gamma rays in relation to dose fractionation and division cells," *Cancer* 34(1974), 39-47.

70. R. A. White and H. D. Thames, Jr., "State-vector models of the cell cycle. II. First three moments of the transit-time distribution," *J. Theoret. Biol.* 77(1979), 141-160.

THEORETICAL AND EXPERIMENTAL INVESTIGATIONS OF MICROBIAL COMPETITION IN CONTINUOUS CULTURE

Paul Waltman

Department of Mathematics
Arizona State University
Tempe, Arizona

Stephen P. Hubbell

Department of Zoology
University of Iowa
Iowa City, Iowa

Sze-Bi Hsu

Department of Mathematics
University of Utah
Salt Lake City, Utah

SHORT HISTORY OF THE CLASSICAL THEORY OF ECOLOGICAL COMPETITION

The classical theory of ecological competition between two or more species, attributed to Lotka and Volterra [60], is an extension of the basic logistic model of single-species growth that dates from Verhulst [59]. The dynamical equations for this theory for two competitors, 1 and 2, are often written as:

$$\frac{dN_1}{dt} = r_1N_1\left[1 - \frac{N_1 + \alpha N_2}{K_1}\right]$$

$$\frac{dN_2}{dt} = r_2N_2\left[1 - \frac{\beta N_1 + N_2}{K_2}\right]$$

Paul Waltman's current affiliation:
Department of Mathematics, University of Iowa, Iowa City, Iowa.

Sze-Bi Hsu's current affiliation:
Department of Applied Mathematics, National Chiao Tung University, Hsinchu, Taiwan.

where N_i is the number of the ith competing species, r_i and K_i are the intrinsic rate of increase and the carrying capacity of the ith competitor, respectively, and α and β are the interaction or "competition" coefficients, expressing the per capita competitive effects of species 2 on 1, and 1 on 2, on the growth rate and realized carrying capacity of the rival species. In the absence of competition ($\alpha = \beta = 0$), each population grows to its respective carrying capacity. In the presence of competition, one or the other rival may survive while its competitor dies out, or else the rivals may coexist. In the two-species case, there are four possible outcomes provided that the initial populations are both positive; which outcome occurs depends on the carrying capacities and competition coefficients. Competitive stability (coexistence) occurs when $\alpha < K_1/K_2$ and $\beta < K_2/K_1$, competitive instability (initial number of each rival determines winner) occurs when these inequalities are both reversed, and competitive dominance (one or the other species wins regardless of initial numbers) occurs when one but not both of these inequalities are reversed.

This classical theory of competition and its extension to n competing species has been the subject of a great amount of theoretical (cf. review, [61, 17]) and experimental [23, 37, 40, 44, 57, 63] work in the last 40 years. However, in recent years there has arisen a widespread feeling that the subject of competition is ready for a new theoretical framework. A pervasive problem with classical theory is that it is "phenomenological," seeking to describe how the numbers of competitors change without ever being specific about which resources are the focus of competition, nor about how efficiently the rivals exploit or control these limited resources. While the lasting appeal of Lotka-Volterra theory has come from its generality and simplicity, this same generality has also made it very difficult for the experimentalist to measure the theory's critical parameters. It has proved especially difficult to estimate the competition coefficients independently of actually growing the competing species together. Usually they have been estimated by fitting the equations to the growth curves of the species in competition (e.g., [57]). Whenever these coefficients can only be estimated from the dynamics of populations already in competition, the value of the theory for prediction is much diminished. It then becomes at best an *ex post facto* description of the outcome of competition [47, 61], and at worst an unsuccessful exercise in curve fitting.

There are a number of other problems with the classical theory which concern its biological assumptions. These include the assumptions of a constant carrying capacity, ecological equivalence of all individuals within each population (no age-dependent differences in birth or death rates or in resource use, for example), no time lags, and constant, linear per-capita effects on population growth rates within and between species. The organisms which best meet these assumptions in general are microorganisms, which as single-celled organisms usually reproduce by simple binary fission, producing clones of genetically identical daughter cells. It is not surprising, therefore, that the experimental work best supporting the theory has been done on microorganisms, beginning with the pioneering work of Gause [7] and continuing until quite recently [8, 40, 56]. In the work with metazoans, however, the theory has with a few possible exceptions (e.g., [23]) not proved adequate [37, 44, 48, 49, 62].

During the last 20 years increasing attention has been given to the details of the processes underlying consumer-resource interactions, with the goal of constructing more mechanistic theories of interspecific competition. Research has been focused on three principal questions. First of all, do the rival species compete only indirectly by lowering the shared pool of limited resources (exploitative competition), or do they also compete more directly by harming their rivals or by sequestering some of the resources for their exclusive use (interference competition) [33]? Secondly, how efficiently do the rivals exploit these limiting resources? In particular, how do the per capita consumption rates of each species respond to a change in resource concentration (nutrients, prey, etc.) in the environment ("functional" response)? Finally, how do these resources, once consumed, translate into a particular rate of population growth ("numerical" response)? There are other questions as well, such as competition within and between age-structured populations (e.g., [39]), and genetic considerations (e.g., [41]), that as yet have not received much attention.

In this chapter we focus exclusively on *exploitative competition*. Interference competition, while common in nature, is mediated through a diversity of mechanisms. As yet there is little consensus about the way that interference should be modeled mathematically given that the effects of toxins or injury are so varied, and are different from the effects of resource sequestering. In any event, it is not a trivial exercise to consider the consequences of exploitative competition, which occurs more universally than interference. Unless one species can totally exclude its rivals from access to the limiting resources, consumption of these resources by both species occurs, and exploitative competition is a reality.

Moreover, there is greater agreement on the biology and mathematics of exploitative competition. Extensive studies have been made on the functional response of a great variety of organisms to resource density, including microorganisms (e.g., [9]), protozoa [46], insect predators [10, 15] and parasitoids [10, 11], fish [24, 36], birds [45, 56], and mammals [13], and others. Functional responses of all organisms are saturating functions of increasing resource density, such that the consumption rate reaches some maximum at high resource density. At low resource density, consumption rate may increase in a nearly linear fashion with resource density (typical of filter-feeding organisms consuming prey much smaller than themselves; e.g., clams, baleen whales); consumption rate may increase nonlinearly, decelerating smoothly to a maximum feeding rate asymptote (most invertebrate predators and many vertebrate predators feeding on one prey type at a time); or consumption rate may increase slowly at low resource density and faster at higher resource density in an S-shaped curve (typical predators that develop learned "search images" as a function of prey encounter rate, and which actively switch between alternate prey, or between non-feeding and feeding behavior, as some threshold prey density). These classes of functional responses have been classified by Holling [14] as Types I, II, and III, respectively. Type II is the most common type of functional response among microorganisms and small invertebrates. In microorganisms, resource uptake occurs at the level of enzyme-mediated transport of specific nutrients across the cell wall, and uptake rates are generally characterized by the Michaelis-Menten equations for enzyme-catalyzed reactions [6, 42]. Types II and III functional responses in higher organisms follow identical mathematics, as has been explored in some detail by Real [43].

Once the limiting resources have been consumed, they may interact in a variety of ways to promote population growth. Leon and Tumpson [30] have distinguished two important classes or resources: complementary and substitutable. Complementary resources are substances which are metabolically independent requirements for growth, such as a carbon and a nitrogen source for a bacterium, or silica and phosphorus for a diatom. Substitutable resources are substances which are metabolically interchangeable in the organism, such as two carbon sources, or two sources of phosphorus. In the case of growth limited by complementary resource, only one such resource can be limiting growth at any given time, and which complementary resource is limiting is determined by the relative rate of supply of the resources in relation to the required proportional demand of the organism. On the other hand, in the case of substitutable resources, growth on any one of the resources is possible because each substitutable resource is actually an alternative form of supply of the same basic nutrient for which there is a requirement. Thus, a given complementary resource can be supplied in a variety of substitutable "packages." For example, planktonic algae obtain their phosphorus from both inorganic sources such as orthophosphate as well as from dissolved organic molecules containing phosphorus. Resources may also be "imperfectly substitutable" if they can be interconverted by the organism to meet various metabolic demands but only by augmenting energy expenditure and generally at the expense of a reduced growth rate.

The developing theories of resource-based ecological competition will, almost certainly, first be tested in the laboratory with systems of competing microorganisms and protozoa. The advantages of the laboratory environment for controlling extraneous variables are clear. Furthermore, microorganisms offer the advantages of (a) rapid generation time, so that experiments can be carried to completion in a short time; (b) small size, so that competitive communities can be economically housed and replicated; (c) clonal homogeneity of individuals, so that genetic differences among individuals within species, barring mutation, are absent; (d) reproduction by binary fission, so that ecological differences due to size and age are minimal; and (e) simple functional responses to resource density, so that the complex behavioral patterns of higher predators (such as searching images, learning, switching behavior, etc.) can be initially ignored.

Consequently, in this chapter we develop the theory of exploitative competition for microbial organisms competing in mixed-growth laboratory cultures. The theory is developed for a culturing technique known as "continuous culture," the most widely used laboratory idealization of a constant carrying capacity environment. The remaining sections of the chapter are organized as follows: First, the technique of continuously culturing microorganisms is briefly described, followed by a section detailing the mathematical model of single-species and multiple-species growth in continuous culture on a single limiting resource. The next section presents the mathematical analysis of the n-species, 1-resource model. This is followed by two sections which give some experimental results of tests of the model, and then generalize by analogy to competition among unicellular planktonic algae in lakes and oceans. Next we consider what happens when two competitors are predators feeding on the same population of prey, a different situation insofar as the "resource" is now capable of self-renewal. We conclude with a brief look at some of our theoretical work in progress.

ORIGINS OF THE CONTINUOUS CULTURE TECHNIQUE

Most natural environments are inhabited by a great diversity of interacting microorganisms, but the ecology of these organisms and their competitive, mutualistic, or predator-prey relationships are almost always very difficult to study in nature because of the highly complex structure of natural environments and because of the limitations in ability to make accurate estimates of natural population densities. In order to study these organisms and their interactions in any detail, it was necessary to develop various isolation procedures so that individual species or strains of microorganisms could be cultured separately in the laboratory. At present laboratory cultures exist for literally thousands of bacteria, fungi, protozoa, and unicellular planktonic and benthic algae, from environments as diverse as lakes, streams, oceans, hot springs, soil, root nodules of plants, and the intestinal tracts of man and a host of other animals.

Many of the pure cultures of these organisms were isolated from so-called "batch enrichment" cultures, in which a small sample of the environmental medium or substrate is incubated with an enriched mixture of nutrients in a closed culture vessel. One or more of the microorganisms present in the sample grow to very large numbers under such conditions, making it easier to isolate individual cells of the organisms into pure culture. These batch cultures are also widely used in studies on the energy metabolism and nutrient requirements of different species of microorganisms.

From an ecological perspective, however, the batch culture environment has some serious drawbacks as a model of natural microbial environments [25, 51, 58]. For one thing, in nature microorganisms almost never encounter the very high nutrient levels that characterize batch cultures. Generally the concentrations of limiting nutrients in soils and natural waters are several orders of magnitude lower than in batch culture (although this is less the case for intestinal environments). Thus, the question arises as to whether the species isolated from batch culture are likely to be representative of the microorganisms that are of a major functional importance in the natural ecosystem. Indeed, one of the significant points of this chapter is that there are both theoretical and experimental grounds for expecting profound differences between the organisms which become abundant under nutrient-rich, batch-culture conditions, and those which are abundant under the much lower nutrient conditions in nature. In batch culture there is also the problem that nutrient conditions are in continual flux with the possibility that growth is not always limited by the same nutrient. This can make the study of the dependence of growth rate on nutrient concentration very difficult, and can obscure the competitive relationships among organisms growing in mixed culture.

The concept of a "continuous" culture was introduced in the late 1940s, and came into widespread application in the 1950s. Continuous cultures were mainly developed so that microbial growth could be studied under nutrient limitation in a controlled nutrient environment. The elementary design and theory of continuous culture was first described by Monod [35] and independently by Novick and Szilard [38], who called their culture device a "chemostat."

The basic concept of a chemostat is to supply the culture continuously with a constant input of sterile medium, and to remove medium plus cells and byproducts from the culture at the same rate, maintaining culture volume constant. Initially the culture is inoculated with a small number of cells; and these multiply until a steady-state cell density is achieved. The influent medium provides all nutrients essential for growth in excess of demand except for one, which is supplied in growth-limiting amounts. Herein lies the principal advantage of continuous culture over batch culture: the rate of dilution controls the rate of microbial growth via the concentration of the growth-limiting nutrient in the medium. As long as the dilution rate is lower than the maximum growth rate attainable by the microorganisms, the cell density will grow to a point at which the cell division rate ("birth" rate) exactly balances the cell washout rate ("death" rate). This steady-state cell density is characterized by a constancy of all metabolic and growth parameters. On the other hand, if the dilution rate exceeds the maximum cell division rate, then total washout of the entire cell population occurs.

The dependence of microbial growth rate on the concentration of limiting nutrient was originally described by Monod [34] as simply a data-fitting curve. Later the relationship was frequently interpreted in terms of Michaelis-Menton kinetics. This explanation is undoubtedly too simplistic for most growth-limiting substances in that the model of a single enzyme-mediated reaction as the one rate-limiting step may not be completely accurate. It now appears that some deviations from the Monod relationship in microorganisms may depend on whether the limiting substrate is an energy source or is an essential nutrient such as a vitamin or mineral [25].

The most common deviation from the classical Monod formulation is that the yield coefficient is not independent of growth rate. This is usually due to the fact that cell volume in many microbial species is a function of steady-state growth (division) rate, and therefore depends on the particular dilution rate that is used in the continuous culture experiment [25, 58]. Other deviations may occur if growth rate falls to zero at some nonzero concentration of the limiting substrate, producing a threshold phenomenon in growth, or it may be found that growth on one substrate is inhibited by the presence of a second substrate. Bacteria, for example, commonly exhibit "diauxic" growth in the presence of two sugars as energy and carbon sources. The presence of glucose in the medium completely inhibits the uptake and metabolism of lactose in *Escherichia coli* via the lac operon [1], until the supply of glucose in the medium is exhausted.

Finally, the cells may be capable of "luxury" consumption of certain nutrients, such that there is uptake and storage of the nutrient in excess of the amounts that are currently needed for growth. This can lead at least to transient departures from the Monod growth rate predictions because growth rate is no longer a strict function of external concentration. Until steady state is reached, growth rates may remain higher than predicted from the Monod relationship or from the concentration of the external nutrient pool. For example, many planktonic algae are capable of considerable luxury consumption of orthophosphate, stored as polyphosphate in the cells; and elevated growth rates can be maintained on these internal stores for some time after external phosphate concentrations have fallen virtually to zero [5].

In spite of departures in detail from the original microbial growth model proposed by Monod, this model remains the simplest and the most widely applicable theory for nutrient-limited growth in microorganisms. Moreover, the theory, as extended to cover n-species microbial competition, now appears sufficient to make qualitatively accurate predictions of the outcomes of microbial competition in continuous culture. For example, Tilman [52], in studying the ability of the Monod model to predict the winning species among diatoms competing for silica and phosphorus, found that the Monod model did as well or better than a more complex "internal stores" model developed by Droop [5], which allows for luxury consumption and uncouples growth rate from external phosphorus concentrations. We also report in the present article and elsewhere [9] that the Monod formulation successfully predicts the outcome of competition between auxotrophic strains of bacteria grown in continuous culture.

Consequently, we feel that competition theory developed from the microbial growth equations of Monod remains the most general theory available, and will serve as the most practical tool for probing the mechanics of microbial competition for some time to come. We also feel that this theory will serve as the foundation for the development of much more mechanistic theories of competition among higher organisms as well, theories which make explicit the interaction of the competitors with each other and with their shared, limited resources.

DERIVATION OF THE MODEL EQUATIONS

The equations for the chemostat for one population were originally derived by Monod [35]. Here we give a simpler derivation following Herbert, Elsworth, and Telling [12], but based on Monod's observations. Let x(t) denote the concentration of the organism and S(t) the concentration of the substrate at time t.

If the organism were grown in a batch culture then the rate of consumption of the substrate and rate of growth of the organism are directly proportional (Monod, [34]):

$$\begin{Bmatrix}\text{rate of growth}\\ \text{of organism}\end{Bmatrix} = y\begin{Bmatrix}\text{rate of consumption}\\ \text{of resource}\end{Bmatrix} \tag{1}$$

y is called the yield constant and is determinable over a finite period of time by

$$y = \frac{\text{weight of the organism formed}}{\text{weight of the substrated used}}$$

The rate of growth of the organism may be simply expressed as

increase = growth - output

or

$$\frac{dx}{dt} = \mu x - Dx \tag{2}$$

where μ is a function (defined below) and D is a constant. The change in the substrate is slightly more complicated in that

increase = input - output - consumption

or

$$\frac{dS}{dt} = DS^{(0)} - DS - \frac{\mu x}{y} \tag{3}$$

where Eq. (1) has been used to model the consumption (keeping in mind that the rate of growth in the concentration of the organism is μx). Finally, μ is assumed (or known experimentally, Monod [34]) to have the form

$$\mu = m \frac{S}{a + S}$$

where m is the maximum growth rate and a is the half-saturation, or Michaelis-Menton, constant, numerically equal to the substrate concentration at which $\mu = m/2$. Combining the above yields the equations of the chemostat:

$$S' = (S^{(0)} - S)D - \frac{1}{y}\frac{mxS}{a + S}$$

$$x' = \frac{mxS}{a + S} - Dx \tag{4}$$

$$x(0) > 0, \qquad S(0) > 0$$

Taylor and Williams [50] extended the derivation to cover n populations existing on one resource, obtaining

$$S' = (S^{(0)} - S)D - \sum_{i=1}^{n} \frac{1}{y_i}\frac{m_i x_i S}{a_i + S}$$

$$x_i' = \frac{m_i x_i S}{a_i + S} - Dx_i \qquad (5)_n$$

$$x_i(0) = x_{i0} > 0, \qquad S(0) = S_0 > 0, \qquad i = 1, \ldots, n$$

ANALYSIS OF THE CHEMOSTAT

The system $(5)_n$ was investigated numerically by Taylor and Williams [50] who found that only one population survived. To describe which one, let $b_i = m_i/D$ and if $b_i > 1$, let $\lambda_i = a_i/(b_i - 1)$. If $b_i > 1$, $i = 1,\ldots, n$, Taylor and Williams found that one the population with the smallest value of λ_i survived. A mathematical proof of this result was given by Hsu, Hubbell, and Waltman [18], and a much shorter proof given by Hsu [16]. Before describing these results precisely we note first that the model is biologically reasonable.

THEOREM 1 The solutions $S(t)$, $x_i(t)$, $i = 1,\ldots, n$ are positive and bounded.

Proof The positivity of the x_i's follows from the uniqueness of solutions of initial value problems and the fact that each $x_i = 0$ face is invariant under the flow given by $(5)_n$. The positivity of $S(t)$ follows from the inequality

$$S(t) > S(0)\exp \int_0^t \left[\left\{ - D - \sum_{i=1}^{n} \frac{m_i}{y_i} \frac{x_i(\xi)}{a_i + S(\xi)} \right\} d\xi \right]$$

and the boundedness of solutions from the relation

$$S(t) + \sum_{i=1}^{n} \frac{x_i(t)}{y_i} = A_n e^{-Dt} + S^{(0)} \tag{6}$$

where

$$A_n = S_0 + \sum_{i=1}^{n} \frac{x_{i0}}{y_i} - S^{(0)}$$

Eq. (6) is obtained by forming a linear differential equation for the quantity

$$S(t) + \sum_{i=1}^{n} \frac{x_i(t)}{y_i}$$

Next it is convenient to eliminate "inadequate" competitors.

THEOREM 2 If

(i) $b_i \leq 1$,

or

(ii) $\lambda_i > S^{(0)}$ if $b_i > 1$,

then $\lim_{t\to\infty} x_i(t) = 0$.

This theorem states that if the maximum growth rate m_i of the ith organism is less than or equal to the dilution rate D or if the parameter $a_i/(b_i - 1) > S^{(0)}$, the organism will become extinct in the culture. Note that the resulting behavior is competition-independent and reflects excessive dilution or, given the dilution rate, the inability to uptake sufficient nutrient. Since $(5)_n$ is a dynamical system, analyzing $(5)_n$ with the ith equation eliminated (analyzing an appropriate $(5)_{n-1}$) is equivalent to studying the omega limit set of the original $(5)_n$ with $\lim_{t\to\infty} x_i(t) = 0$.

Proof First observe that from (6) if $\varepsilon > 0$ there exists a t_0 such that if $t \geq t_0$, $S(t) \leq S^{(0)} + \varepsilon$. $x_i(t)$ may be written

$$x_i(t) = x_{i0} \exp \int_0^t \frac{(m_i - D)S(\xi) - a_i D}{a_i + S(\xi)} \, d\xi \tag{7}$$

If $b_i \leq 1$, then

$$x_i(t) \leq x_{i0} \exp \int_0^t \frac{-a_i D}{a_i + S(\xi)} \, d\xi$$

$$\leq C x_{i0} \exp \frac{-a_i D}{a_i + S^{(0)} + 1} (t - t_0)$$

where t_0 is chosen so that for $t \geq t_0$, $S(t) \leq S^0 + 1$ and

$$C = \exp \int_0^{t_0} \frac{-a_i D}{a_i + S(\xi)} \, d\xi$$

Since the exponent is negative and $x_i(t) > 0$, $\lim_{t\to\infty} x_i(t) = 0$.

Rearranging (7) yields

$$x_i(t) = x_{i0}\exp \int_0^t \frac{m_i - D}{a_i + S(\xi)} \left[S(\xi) - \frac{a_i}{b_i - 1} \right] d\xi \tag{8}$$

If $b_i > 1$, then the first factor of the integrand is positive. Let $0 < \xi < (a_i/(b_i - 1)) - S^{(0)}$, and choose $t_0 > 0$ such that $S(t) \leq S^{(0)} + \varepsilon$ for $t \geq t_0$. Then for an appropriate constant C, it follows that

$$x_i(t) \leq Cx_{i0}\exp\left\{\left[S^{(0)} + \varepsilon - \frac{a_i}{b_i - 1} \right]\left[\frac{m_i - D}{a_i + S^{(0)} + 1} \right](t - t_0)\right\}$$

The first factor in the exponent is negative, and the other two positive so $\lim_{t\to\infty} x_i(t) = 0$.

For $b_i > 1$, as noted above, we define $\lambda_i = a_i/(b_i - 1)$. The basic hypothesis is

$$\begin{aligned} &\lambda_1 < S^{(0)} \\ &0 < \lambda_1 < \lambda_2 \leq \lambda_3 \leq \cdots \leq \lambda_n \end{aligned} \tag{H$_n$}$$

The equations may be relabeled without loss of generality, so that the parameters $\lambda_i = a_i/(b_i - 1)$ are nondecreasing in i. $(H)_n$ excludes equality of this parameter for the first and any other population.

THEOREM 3 Let $(H)_n$ hold. The solutions of $(5)_n$ satisfy

$$\lim_{t\to\infty} S(t) = \lambda_1,$$

$$\lim_{t\to\infty} x_1(t) = x_1^* = y_1(S^{(0)} - \lambda_1)$$

$$\lim_{t\to\infty} x_i(t) = 0, \qquad 2 \leq i \leq n$$

This theorem states that under the hypothesis $(H)_n$ only one type of organism survives, the one with the lowest value of λ, and gives the limiting concentrations. For a given system, the parameter λ depends on two characteristics, the growth rate and the Michaelis-Menten constant. It is biologically reasonable to assume that for two distinct species, the corresponding parameter will be different. Hence $(H)_n$ (with all strict inequalities) is a biologically reasonable assumption.

Proof of Theorem 3 (Hsu, [16]) Let F denote the positive cone in R^{n+1} and define there a Liapunov function

$$V(S, x_1,\dots, x_n)$$

$$= S - \lambda_1 - \lambda_1 \ln\left(\frac{S}{\lambda_1}\right) + c_1\left[(x_1 - x_1^* - x_1^*)\ln\left(\frac{x_1}{x_1^*}\right)\right] + \sum_{i=2}^{n} c_i x_i$$

where $c_i = m_i/[y_i(m_i - D)]$. Along solutions of the equation

$$\frac{d}{dt} V(S(t), x_1(t),\dots, x_n(t))$$

$$= \text{grad } V \cdot (S', x_1',\dots, x_n')^T$$

$$= \begin{bmatrix} 1 - \dfrac{\lambda_1}{S} \\ c_1\left(1 - \dfrac{x_1}{x_1}\right) \\ c_2 \\ \cdot \\ \cdot \\ \cdot \\ c_n \end{bmatrix} \begin{bmatrix} (S^{(0)} - S)D - \displaystyle\sum_{i=1}^{n} \frac{m_i}{y_i} \frac{x_i S}{a_i + S} \\ \dfrac{(m_1 - D)}{a_1 + S}(S - \lambda_1)x_1 \\ \dfrac{m_2 - D}{a_2 + S}(S - \lambda_2)x_2 \\ \cdot \\ \cdot \\ \cdot \\ \dfrac{m_n - D}{a_n + S}(S - \lambda_n)x_n \end{bmatrix}$$

$$= (S - \lambda_1)\left[\frac{(S^{(0)} - S)}{S} D - \frac{k_1 x_1^*}{a_1 + S}\right] + \sum_{i=2}^{n} \frac{m_i}{y_i}(\lambda_1 - \lambda_i)\frac{x_i}{a_i + S}$$

where $k_1 = m_1/y_1$.

Since

$$x_1^* = y_1(S^{(0)} - \lambda_1)$$

$$= \frac{(S^{(0)} - \lambda_1)(a_1 + \lambda_1)Dy_1}{m_1\lambda_1}$$

then

$$\frac{k_1x_1^*}{a_1 + S} = \frac{k_1\lambda_1x_1^*}{\lambda_1(a_1 + S)} = \frac{(S^{(0)} - \lambda_1)(a_1 + \lambda_1)D}{\lambda_1(a_1 + S)}$$

Thus

$$\frac{(S^{(0)} - S)D}{S} - \frac{k_1x_1^*}{a_1 + S}$$

$$= \frac{D[S^{(0)}\lambda_1a_1 + S^{(0)}\lambda_1S - (S^{(0)} - \lambda_1)(a_1 + \lambda_1)S - \lambda_1S^2] - \lambda_1a_1S}{S(a_1 + S)\lambda_1}$$

$$= \frac{-D(S - \lambda_1)(\lambda_1S + a_1S^{(0)})}{S(a_1 + S)\lambda_1}$$

Using this in the computation of dV/dt yields

$$\frac{dV}{dt} = \frac{-D(S - \lambda_1)^2(\lambda_1S + a_1S^{(0)})}{(a_1 + S)S\lambda_1} = \sum_{i=2}^{n} k_i(\lambda_1 - \lambda_i)\frac{x_i}{a_i + S} \leq 0$$

since $0 < \lambda_1 < \lambda_i$, $i \geq 2$, and $S > 0$. The set $E = \{(S, x_1, \ldots, x_n) | \dot{V} = 0\}$ is given by

$$E = \{\lambda_1, x_1, 0, 0, \ldots, 0)\}$$

Since $\lambda_1 < S^{(0)}$, the only invariant set in E is

$$S = \lambda_1$$

$$x_1 = x_1^* = y_1(S^{(0)} - \lambda_1)$$

$$x_i = 0, \qquad i = 2, \ldots, n$$

An application of LaSalle [29, p. 30] completes the proof. The proof in [16] is more general in that it allows unequal death rates.

Only one coexistence result was obtained.

THEOREM 4 Let $b_1 > 1$ and $0 < \lambda_1 = \lambda_2 = \ldots = \lambda_n < S^{(0)}$. Then

$$\lim_{t\to\infty} S(t) = \lambda_1$$

and

$$\lim_{t\to\infty} x_i(t) = x_i^* > 0$$

where

$$\lambda_1 + \sum_{i=1}^{n} \frac{x_i^*}{y_i} = S^{(0)}$$

EXPERIMENTAL RESULTS

Theorem 3 makes an explicit prediction when several populations are grown in a chemostat--it predicts a unique surviving population and the steady state concentration levels for the nutrient and for the survivor. The experiments corresponding to this prediction have been made by Hansen and Hubbell [9] and will be summarized here. Before giving the details, however, we pause to observe several features of the model. First of all, two of the parameters, $S^{(0)}$, the input concentration, and D, the dilution parameter, are controlled by the experimenter. The yield constants, y_i, are measured by growing the organisms one at a time in batch culture, or (preferably) in continuous culture, and estimating cell concentrations at steady state for a given input of limiting nutrient. Since each equation for a population x_i contains only the variables x_i and S, the nutrient concentration, the remaining parameters, m_i and a_i can also be obtained by growing each population, in the absence of the other organisms, on the limiting nutrient. The most widely used method of computing m_i and a_i is a graphical technique known as a "Lineweaver-Burk Plot" [31]. While the biological details of such parameter estimation are not of interest to mathematicians, the important point is that the determination of m_i and a_i is *independent* of the other populations. For this reason it is possible to make the measurements of each organism and predict the outcome when the two are mixed together and grown in a chemostat.

The λ criterion for competitive ability is nonobvious and requires experimental verification. It could not have been predicted from classical theories of ecological competition. *A priori* it might have been expected that the winner would always be the organism with the highest affinity and lowest a_i, for the limiting nutrient, or perhaps the organism with the highest intrinsic rate of increase. In the Monod model of competition, the intrinsic rate of increase of organism x_i is given by the difference between the maximal birth rate, m_i, and the death rate, D, and is denoted by r_i. The theory, however, asserts that the critical parameter to competitive success is actually a weighted a_i. This weighted a_i is the parameter λ_i which can be rewritten from the fourth section as $\lambda_i = a_i(D/r_i)$. Thus the biologically interesting prediction can be made that a species may actually lose in competition even with a lower half-saturation constant a_i and thus a higher affinity for the resource, if it also has a lower intrinsic rate of increase r_i than its rivals. The theory also asserts that winning will not depend in any way on the growth efficiency or yield of the competing organisms from the limiting resource.

To make a rigorous test of the λ criterion in continuous culture requires proof that (i) if two organisms have equal r_i's and D's, the organism with the lower a_i wins; that (ii) if two organisms have identical a_i's and D's, the organism with the higher r_i wins; and that (iii) if two organisms have different a_i's and r_i's but in spite of this still have identical λ's, then the organisms should coexist. Hansen and Hubbell [9] have conducted all three of these tests in competition experiments with mutant strains of bacteria which must be supplied with an external source of the amino acid, tryptophan, in order to grow and divide. In the first set of tests, the competing strains were of different species and differed primarily in their a_i values for tryptophan uptake. In the second and third tests, the competing organisms were strains of the same species, and of the same mating type so that conjugation between strains and gene exchange were prevented. In the second set of tests, the strains had identical a_i values but differed considerably in maximal specific growth rates and in r_i's. In the third set, the strains had identical λ's although they differed in both their a_i's and r_i's.

Each set of experiments was conducted in two sequential parts. The first part consisted of measuring the a_i's and m_i's for each bacterial strain grown alone in batch culture on a limiting amount of tryptophan. The values of λ were calculated to predict the outcomes of the subsequent competition experiments. The yield constants were measured in pure-strain continuous cultures at steady state.

The measured parameters for each bacterial strain are shown in Table 4.1, in addition to other run parameters for each experiment. Figure 4.1 shows the result of the first experiment. In this case, *Escherichia coli* C-8 was opposed by *Pseudomonas aerogenosa* PAO-283. The a_i values for tryptophan for these two bacterial species differ by nearly two orders of magnitude, and as a result the λ value for *E. coli* is much smaller than for *P. aerogenosa*. The predicted winner, *E. coli*, actually did win, effectively eliminating *P. aerogenosa* in the space of 60 hr. Note that *P. aerogenosa* lost to *E. coli* in spite of having a higher intrinsic rate of increase, and in spite of having a starting 200:1 numerical advantage. That initial densities as different as these do not influence the outcome can be taken as evidence supporting the view that competition for tryptophan between these two bacteria is purely exploitative.

The first experiment confirms the importance of having high affinity (low a_i) for the limiting resource, but the second and third experiments confirmed that it is a weighted a_i (i.e., λ) which is critical to the outcome. In both of these sets of experiments, competition occurred between two clones of *E. coli*. The strains were artificially selected for a cross pattern of drug resistance (strain 1 resistant to drug a, sensitive to drug b; strain 2 with the reverse pattern). These drugs cause a lowering of cell growth rate--a linear depression of growth rate at low drug concentrations (see Figure 4.2). By adding small amounts of a drug to the culture medium flowing into the chemostat, it was possible to alter the intrinsic rate of increase, r_i, of the sensitive strain while leaving the r_i of the resistant

TABLE 4.1

Uptake and Growth Parameters for Competing Bacterial Strains[a]

Experiment Number	Bacterial Strain	Yield Cells/g	a_i g/l	m_i hr^{-1}	r hr^{-1}	λ g/l
1	C-8[b]	2.5×10^{10}	3.0×10^{-6}	0.81	0.75	2.40×10^{-7}
	PAO283[c]	3.8×10^{10}	3.1×10^{-4}	0.91	0.85	2.19×10^{-5}
2	C-8 $nal^r spec^s$	6.3×10^{10}	1.6×10^{-6}	0.68	0.61	1.98×10^{-7}
	C-8 $nal^s spec^r$	6.2×10^{10}	1.6×10^{-6}	0.96	0.89	1.35×10^{-7}
3[d]	C-8 $nal^r spec^s$	6.3×10^{10}	1.6×10^{-6}	0.68	0.61	1.98×10^{-7}
	C-8 $nal^s spec^r$	6.2×10^{10}	0.9×10^{-6}	0.41	0.34	1.99×10^{-7}

Other Run Parameters

Experiment Number	S_0 g/l	D hr^{-1}	Volume ml
1	1×10^{-4}	6.0×10^{-2}	200
2	5×10^{-6}	7.5×10^{-2}	200
3	5×10^{-6}	7.5×10^{-2}	200

[a]The limiting nutrient is the amino acid, tryptophan, needed by both strains. The superscripts "r" and "s" refer to drug resistance or sensitivity, respectively.

[b]*Escherichia coli*

[c]*Pseudomonas aerogenosa*

[d]0.5 μg/ml nalidixic acid added

Metabolic inhibitors (drugs): nal = nalidixic acid
spec = spectinomycin

strain unaffected. This technique was used in the second set of experiments. In these tests, although both strains had identical half-saturation constants, the strain with the lower intrinsic rate of increase lost (see Figure 5.3). In the final set of experiments, the two *E. coli* strains chosen also differed in their half-saturation constants. By lowering the intrinsic rate of increase of the strain with the lower half-saturation constant with a drug, it was possible to make the λ's of the two strains equal. The result of the thrice-replicated experiment (see Figure 4.4) was coexistence of the competing bacteria, as predicted by the theory [18].

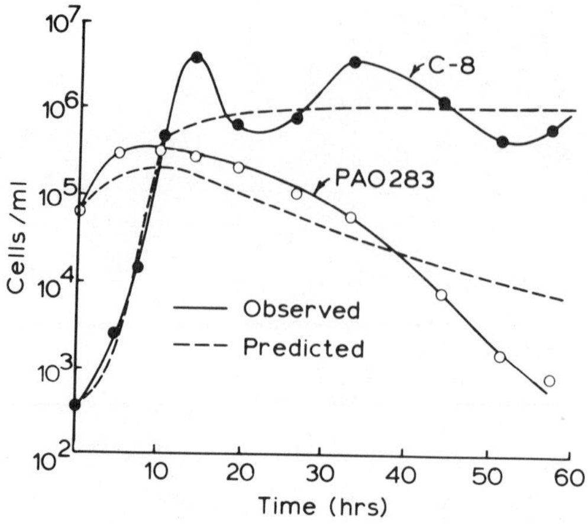

Fig. 4.1. Observed and predicted time course of cell density for the strains PAO283 and C-8 competing in mixed culture for limiting tryptophan in continuous culture. Parameters for the run are listed in Table 4.1. The predicted curves were obtained by numerical solution of equations $(5)_2$. In this experiment, the strains differ principally in their half-saturation constants for tryptophan.

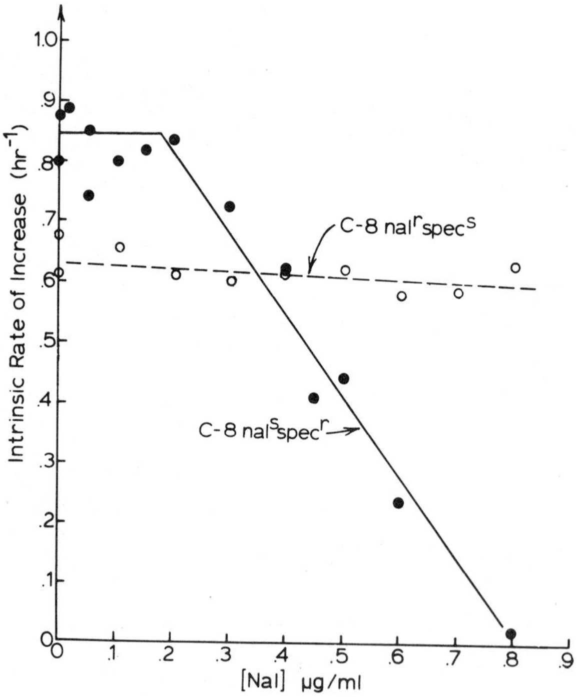

Fig. 4.2. Effect of nalidixic acid concentration on the intrinsic rate of increase of two strains of *E. coli* growing alone. Growth rate of the nal-sensitive strain in linearly depressed with increasing concentration between 0.2 and 0.8 g/ml nalidixic acid. Growth rate of the nal-resistant strain is virtually unaffected. In the figure, read "+" as "resistant" and "-" as "sensitive."

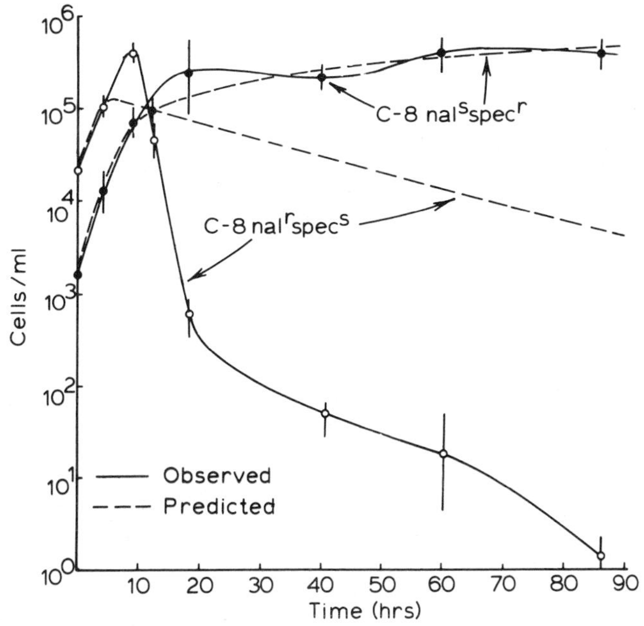

Fig. 4.3. Observed and predicted time course of cell density for two strains of *E. coli* competing for limiting tryptophan. Strains differ in their intrinsic rates of increase, but not in their half-saturation constants (Table 4.1). Dots represent mean cell densities for 3 replicate runs; bars are the ranges of cell densities.

It should be noted that although the qualitative outcomes of the theory are correctly matched by experiment, there are some quantitative deviations of the experimental results from the theoretical trajectories forecast by the system of equations in (5). For example, sometimes oscillations (Figure 4.1) in the approach to steady-state cell density occur in the culture. These could be caused by time delays present in the system that are not reflected in the model. How to formulate the delays so as to give a better fit appears to be an interesting open question. It seems likely that changing cell volume during the growth and plateau phases of population increase could contribute some of the delay. Cell death over and above losses in the effluent could also account for the faster-than-expected decline in the losing strains.

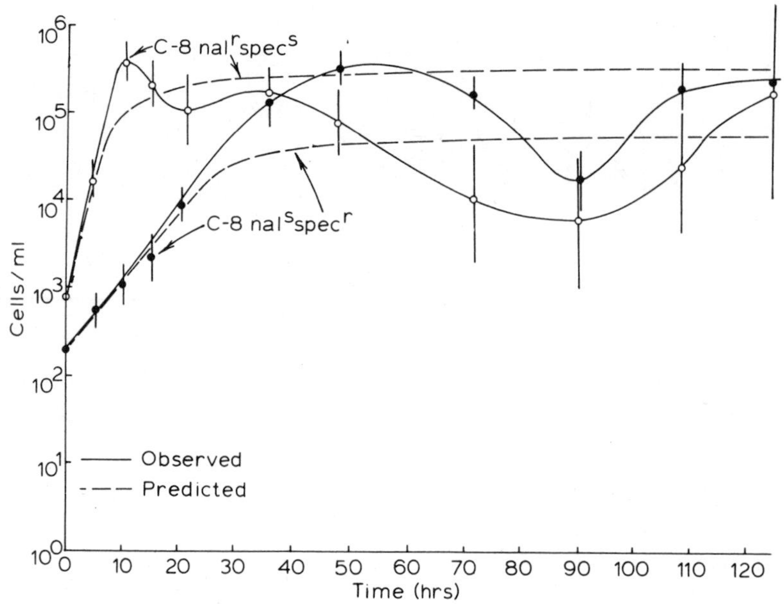

Fig. 4.4. Time course of cell density for two strains of *E. coli* competing for limiting tryptophan in continuous culture. Strains differ in their half-saturation constants and in their intrinsic rates of increase, but are identical in their T values. The strains coexisted for the length of the run.

RELEVANCE TO THE ECOLOGY OF LAKES AND OCEANS

Naturally occurring bodies of water such as lakes and oceans are inhabited by a diverse array of microscopic organisms such as bacteria and unicellular algae that form complex planktonic communities. Because these organisms share many of the same nutrient requirements, the question naturally arises as to the extent to which such ecosystems are analogs of continuous cultures in the laboratory. The surface waters of almost all lakes and oceans, where these planktonic organisms occur, receive nutrient inputs from eroding watersheds or upwelling water rich in nutrients from bottom sediments. Indeed, the planktonic community is wholly dependent on sustained nutrient inputs for continued persistence in the sunlit surface waters [22].

In the temperature zone at least, the biggest departure from the chemostat results from the periodic forcing of natural ecosystems by seasonal weather changes. Lakes and oceans, although in part thermally buffered by the high heat capacity of water, nevertheless do not escape the influence of the seasons. Nutrient levels experienced by phytoplankton change profoundly from one season to the next. For example, in deeper lakes and oceans, gradients of water density established in summer prevent complete mixing of the water column, thereby cutting off the supply of nutrients to surface waters from bottom sediments or nutrient-rich older and denser waters below. In the spring and fall, however, heating and cooling of the surface waters, respectively, equalize the water densities throughout the water column, and more nearly complete mixing occurs. Therefore, these times of the year in temperate bodies of water are characterized by nutrient enrichment of the surface waters. These spring and fall "turnovers" are accompanied by phytoplankton "blooms" for a few weeks during and after turnover. As summer and winter progress, respectively, the nutrient levels decline as a direct result of their consumption and removal by the plankton. The nutrients are slowly lost from the upper waters because the plankton gradually sinks to the bottom, decomposes, and liberates the nutrients once again, to be refluxed to the surface during the next lake overturn. These seasonal changes in nutrient and phytoplankton levels in the surface waters of lakes and oceans are well documented in the limnological literature [22, 26].

In addition to these almost periodic seasonal drivers, natural "cultures" of microorganisms, especially smaller bodies of water, receive stochastic nutrient pulses of varying amplitude and duration. These pulses are caused by nutrient input during precipitation and the resulting, fairly abrupt, increased inflow from the surrounding watershed. Pulses of this type have been measured for a number of important limiting nutrients on several occasions [26].

In spite of the fact that natural planktonic ecosystems often depart from the ideal laboratory chemostat, both in the inconstancy of nutrient input and in the lack of achievement of a steady state, nevertheless a number of predictions about the composition of these communities based on chemostat studies have been very successful. This has been particularly true among planktonic algal communities. The success of these predictions appears due to the fact that, in spite of natural fluctuations in nutrients, the same nutrients remain limiting to the plankton throughout, or else quickly become limiting once again after the pulse is over.

These successes can be illustrated by recent work on freshwater diatoms. These unicellular algae are encased in a siliceous shell, which imposes on all diatoms a requirement for silica. Silicate has a fairly low solubility in water, and as a result diatoms are frequently limited by the availability of this mineral. Kilham [27] was able to relate the dominance of particular diatom species in a series of African lakes to the concentration of silicate in the streamwater flowing into the lake, and to the amount of silica in the bedrock of the watershed. Diatom species with lower half-saturation constants for uptake of silicate were dominant in lakes with lower input concentrations of silicate. Kilham also found, however, that lakes with a higher silicate input were dominated with diatom species having higher half-saturation constants for silicate. This result would not be expected from chemostat theory if all diatoms in these lakes were limited by silicate.

Later work by Tilman and Kilham [53] and Tilman [52, 55] suggests a very simple explanation: limitation by different complementary nutrients in the different lakes. They studied two species of diatom, *Asterionella* and *Cyclotella*, growing first under silicate limitation and then under phosphate limitation. *Cyclotella* not only has the lower half-saturation constant, but also the higher intrinsic rate of increase, under silicate limitation, and the converse is true under phosphate limitation. Therefore, when these diatoms are placed in competition in mixed-growth continuous culture, *Cyclotella* wins when both species are silicate-limited, whereas *Asterionella* wins when both species are phosphate-limited. Tilman also found a region of coexistence, corresponding to nutrient ratios of phosphate to silicate in the influent medium for which *Asterionella* was silicate-limited but *Cyclotella* was phosphate-limited. These results were fully predicted by chemostat theory extended to two complementary resources.

These two diatom species are abundant in the Great Lakes, and Tilman [52] went on to show that chemostat theory could predict the relative abundance of these species with reasonable accuracy. To make these predictions he assumed that the sampled lake waters were at or near steady state. Then the observed phosphate/silicate ratios in the water samples were used to predict the relative proportion of *Cyclotella* cells in the water sample. When this ratio is high, generally in inshore waters where phosphate levels are at their highest, the predicted percentage of *Cyclotella* was zero, due to exclusion by *Asterionella*. In midlake water samples, where phosphate levels are at their lowest, the expected percentage of *Cyclotella* was 100%, resulting from the competitive elimination of *Asterionella*. In the region of intermediate phosphate/silicate ratios, the theory predicted intermediate percentages of *Cyclotella*. The fit of the expected percentages to the observed percentages of *Cyclotella* was remarkably good: more than 70% of the variance was explained. This is a very high percentage of variance explained, given that many other variables also affect diatom abundance in nature, and the further fact that steady-state conditions were assumed in order to make the predictions.

As mentioned in the second section, Tilman also discovered that a model of luxury consumption of phosphate by the diatoms, which takes into account internal stores of phosphate in the cells, generally fit the data no better than the simpler Monod model. In any event, at steady state the predictions were virtually identical for the Monod formulation and the "internal stores model" proposed by Droop [5] for unicellular algae. Recently, it has been shown that there is a good theoretical basis for steady-state equivalence between these two models [3].

These studies have been presented for illustrative purposes to show how well the theory is being adapted to natural ecosystems. Because the theory and its biological testing are still in their infancy, we can expect many more theoretical and biological contributions to resource-based competition theory in the near future.

TWO COMPETITORS AND A SELF REVIEWING RESOURCE

The Equations

The chemostat has forced resource input, but in many ecological systems the resource is not a chemical but a reproducing organism. Thus it is desirable to change the system (5) to reflect this phenomenon. However, doing so introduces serious mathematical complications. The simplest model of limited population growth of a simple organism is the logistic equation. Replacing the chemostat input with logistic terms, keeping the Michaelis-Menten dynamics (here, more appropriately, called Holling dynamics [13]), and limiting n to be 2, yields a system of the form

$$S' = \gamma S\left(1 - \frac{S}{K}\right) - \sum_{i=1}^{2} \frac{m_i}{y_i} \frac{x_i S}{a_i + S}$$

$$x_i' = \frac{m_i S x_i}{a_i + S} - D_i x_i \qquad (9)$$

$$x_i(0) = x_{i0} > 0, \quad S(0) = S_0 > 0, \quad i = 1, 2$$

γ is the intrinsic growth rate for the resource, now appropriately viewed as the prey population, and K is the carrying capacity--the natural limit of the population size without predators--and all of the other constants are as before except that now there is an individual death rate rather than a simple washout rate. The use of D_i reflects the fact that this parameter may be different for each population. We seek to analyze the system (9). Where a rigorous analysis has not been achieved, computer simulation has been utilized to indicate the behavior of the system. The mathematical proofs and a more complete biological discussion can be found in Hsu, Hubbell, and Waltman [19] and [20]. In this section we summarize the basic results of these papers.

Statement of the Mathematical Results

As with $(5)_n$ it follows easily that all solutions with initial conditions in the positive octant are bounded and remain in the positive octant. In the analysis of (9) the carrying capacity K plays the role of the input nutrient $S^{(0)}$ in $(5)_n$ in the sense that if the maximum attainable amount of resource, K, is inadequate, survival is not possible. Also one expects that if the death rate is larger than the maximum possible birth rate, survival is not possible. This is the content of the following:

THEOREM 5 If $b_i \leq 1$ or if $K \leq \lambda_i$, $\lim_{t\to\infty} x_i(t) = 0$.

If one of the above conditions is satisfied, then the behavior of the solutions of (9) is determined from the analysis of the remaining two equations whose solutions form the omega limit set of the entire dynamical system. (Of course if for each i = 1 and 2, one of the above conditions is satisfied, the omega limit set of every trajectory is the critical point (K, 0, 0).) The interesting behavior of the two dimensional system

$$S' = \gamma S\left(1 - \frac{S}{K}\right) - \frac{m}{y}\frac{xS}{a + S}$$

$$x' = \frac{mxS}{a + S} - Dx \tag{10}$$

$$x(0) = x_0 > 0, \qquad S(0) = S_0 > 0$$

is contained in the following statement.

LEMMA 6 If in (10) $K < a + 2\lambda$, where $\lambda = aD/(m - D)$, then (10) has no limit cycles in the first quadrant. If $K > a + 2\lambda$, there exists a limit cycle in the first quadrant.

The proof of the first statement follows from the Dulac Criterion; the second, from boundedness of solutions, the Poincaré-Bendixon Theorem, and the instability of the interior critical point. When one competitor is "inadequate" ($b_i \leq 1$ or $\lambda_i \geq K$), then with relatively little additional effort one obtains:

THEOREM 7 Let (a) $0 < \lambda_1 < K$, and

(b) $\lambda_2 > K$ or $b_2 \leq 1$

If

$$K < a_1 + 2\lambda_1$$

then

$$\lim_{t\to\infty} S(t) = S^* = \lambda_1$$

$$\lim_{t\to\infty} x_1(t) = x_1^* = \frac{\lambda\left(1 - \frac{S^*}{K}\right)(a_1 + S^*)}{(m_1/y_1)}$$

$$\lim_{t\to\infty} x_2(t) = 0$$

If $K > a_1 + 2\lambda_1$, then the omega limit set of the trajectory of $(S(t), x_1(t), x_2(t))$ lies in the $S - x_1$ plane (i.e., $\lim_{t\to\infty} x_2(t) = 0$) and contains a periodic trajectory except for one distinguished orbit which approaches the critical point $(S^*, x_1^*, 0)$.

The interesting case, of course is when both competitors can survive alone on the resource. To have competitive exclusion hold, one seeks conditions which make the omega limit set two dimensional. One such criterion is the following:

LEMMA 8 If $0 < \lambda_1 < \lambda_2$ and if $b_2 \leq b_1$, then $\lim_{t\to\infty} x_2(t) = 0$.

This lemma provides the technical bases for our principal result on competitive exclusion for the system (9).

THEOREM 9 Suppose that $0 < \lambda_1 < \lambda_2 < K$ and $b_1 \geq b_2 > 1$. Then the conclusions of Theorem 7 hold as $K < a_1 + 2\lambda_1$ or $K > a_1 + 2\lambda_1$.

Thus coexistence is possible only if $\lambda_1 < \lambda_2$, $a_1 < a_2$, and $b_1 < b_2$. (Note that these conditions are not independent.) We also have the following result on the persistence of x_1.

THEOREM 10 Suppose that $0 < \lambda_1 < \lambda_2 < K$, $a_1 < a_2$ and $K < a_2 + 2\lambda_2$. Then $\limsup_{t\to\infty} x_1(t) > 0$.

In the numerical simulation, the following result was useful, particularly when $b_2 - b_1$ was small.

THEOREM 11 Suppose that $0 < \lambda_1 < \lambda_2 < K$, $a_1 < b_1$, $b_1 < b_2$, and $K < (b_1 a_2 - b_2 a_1)/(b_2 - b_1)$. Then $\lim_{t\to\infty} x_2(t) = 0$.

Numerical Studies

The preceding theorems yield sufficient conditions for one predator population to survive and the ordering $\lambda_1 < \lambda_2$ favors the first predator. In view of the results presented in the third section one might expect that this is the only outcome possible. Extensive computer simulation shows this not to be the case; in fact, not only may coexistence occur but if K is sufficiently large (relative to other parameters) predator two can win the competition. A coexistence model of two predators which can survive on a single resource have been constructed by McGehee and Armstrong [32] and numerical studies by Koch [28] proceed ours. We summarize our computer results with three graphs from [20].

The data in Figure 4.5 was obtained by fixing all of the parameters except a_1 and K and letting these vary subject to $\lambda_1 < \lambda_2$. The designations

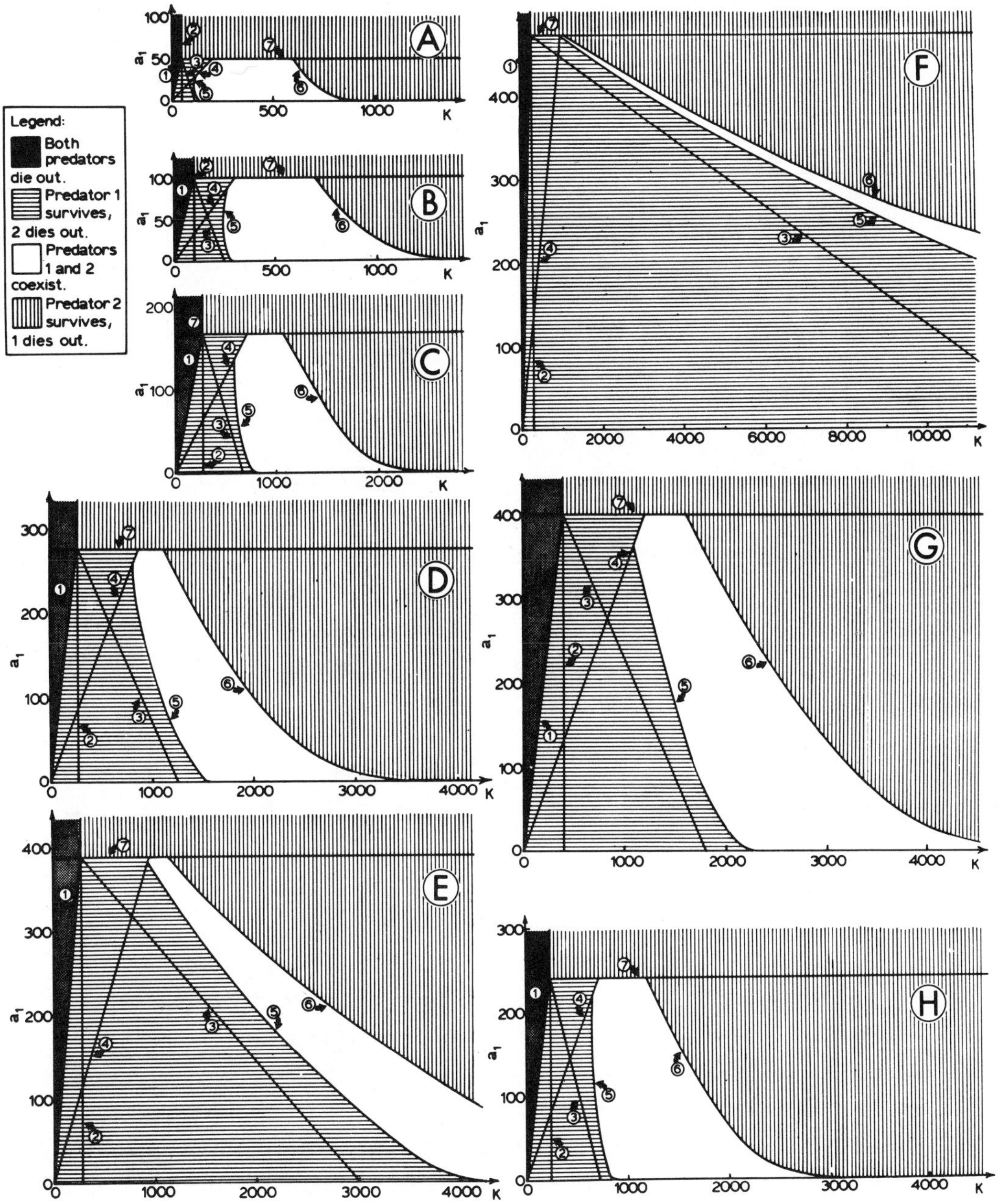

Fig. 4.5. Coexistence region illustrated for 2 predator species competing for a single prey species. Parameter space is a plot of prey carrying capacity, K, on the x-axis against the half-saturation constant for predator 1, a_1, on the y-axis. Parameters fixed for all graphs: $\gamma = 20.\ln 2$; $d_1 = \ln 2/2$; $d_2 = \ln 2$; $y_1 = 0.1$; $y_2 = 1.14$. Parameters for each case: A: $a_2 = 500$, $m_1 = \ln 2$ ($b_1 = 2$), $m_2 = 11 \cdot \ln 2$ ($b_2 = 11$); B: $a_2 = 500$, $m_1 = \ln 2$

($b_1 = 2$), $m_2 = 6 \cdot \ln 2$ ($b_2 = 6$); C: $a_2 = 500$, $M_1 = 0.8 \cdot \ln 2$ ($b_1 = 1.6$), $m_2 = 2.8 \cdot \ln 2$ ($b_2 = 2.8$); D: $a_2 = 500$, $m_1 = \ln 2$ ($b_1 = 2$), $m_2 = 2.8 \cdot \ln 2$ ($b_2 = 2.8$); E: $a_2 = 500$, $m_1 = 1.2 \cdot \ln 2$ ($b_1 = 2.4$); $m_2 = 2.8 \cdot \ln 2$ ($b_2 = 2.8$); F: $a_2 = 500$, $m_1 = 1.35 \cdot \ln 2$ ($b_2 = 2.7$), $m_2 = 2.8 \cdot \ln 2$ ($b_2 = 2.8$); G: $a_2 = 720$, $m_1 = \ln 2$ ($b_1 = 2$), $m_2 = 2.8 \cdot \ln 2$ ($b_2 = 2.8$); H: $a_2 = 720$, $m_1 = \ln 2$ ($b_1 = 2$), $m_2 = 4 \cdot \ln 2$ ($b_2 = 4$). Numbered lines are as follows: (1): $\lambda_1 = K$; (2): $\lambda_2 = K$; (3): $K = (a_2 b_1 - a_1 b_2)/(b_2 - b_1)$; (4): $\lambda_1 = (K - a_1)/2$; (5): Lower $k - a_1$ boundary for coexistence region (known from numerical analysis); (6): Upper $K - a_1$ boundary for coexistence region (known from numerical analysis); (7): $\lambda_1 = \lambda_2$.

A - H indicate different parameters (see the figure caption). The numbered lines 1-4 correspond to analytically known results whereas curves 5 and 6 are numerically determined, and line 7 corresponds to $\lambda_1 = \lambda_2$. The horizontally shaded area to the left of curve 5 and below line 7 corresponds to the (numerically determined) region in $a_1 - K$ space where predator 1 wins the competition. The vertically shaded area to the right of curve 6 and below line 7 corresponds to the region where predator 2 wins the competition. In the unshaded area between curves 5 and 6, coexistence was found to occur as a globally asymptotically stable limit cycle. That the coexistence region is nonempty is a consequence of the work of G. Butler [2]. That the solutions corresponding to parameters in the coexistence region tend to a periodic solution is an open mathematical question. Figure 4.6 shows such a periodic solution and its projection onto $x_1 - x_2$ space.

In Figure 4.7 shows a cut through Figure 4.5H, $a_1 = 100$. We interpret this figure in the language of bifurcation theory. For values of $K < \lambda_1$, the critical point (K, 0, 0) is globally asymptotically stable (no predator survives). As K passes through λ_1, a second critical point enters the positive octant, (K, 0, 0) loses its stability (it has a one dimensional stable manifold) and the new critical point (λ_1, x^*, 0) is globally asymptotically stable. At $K = a_1 + 2\lambda_1$, this critical point bifurcates (a Hopf bifurcation) into a globally asymptotically stable limit cycle, lying in the $S - x_1$ plane and (λ_1, x_1^*, 0) becomes unstable (has a two dimensional unstable manifold). The upper and lower curves in Figure 4.7 show the maximum and minimum of the periodic solution. At a higher value of K--determined only numerically--this limit cycle bifurcates (not a Hopf bifurcation) into two periodic solutions, one remaining in the $S - x_1$ plane and one in the open, positive octant. This is the region of coexistence. For even higher values of K,

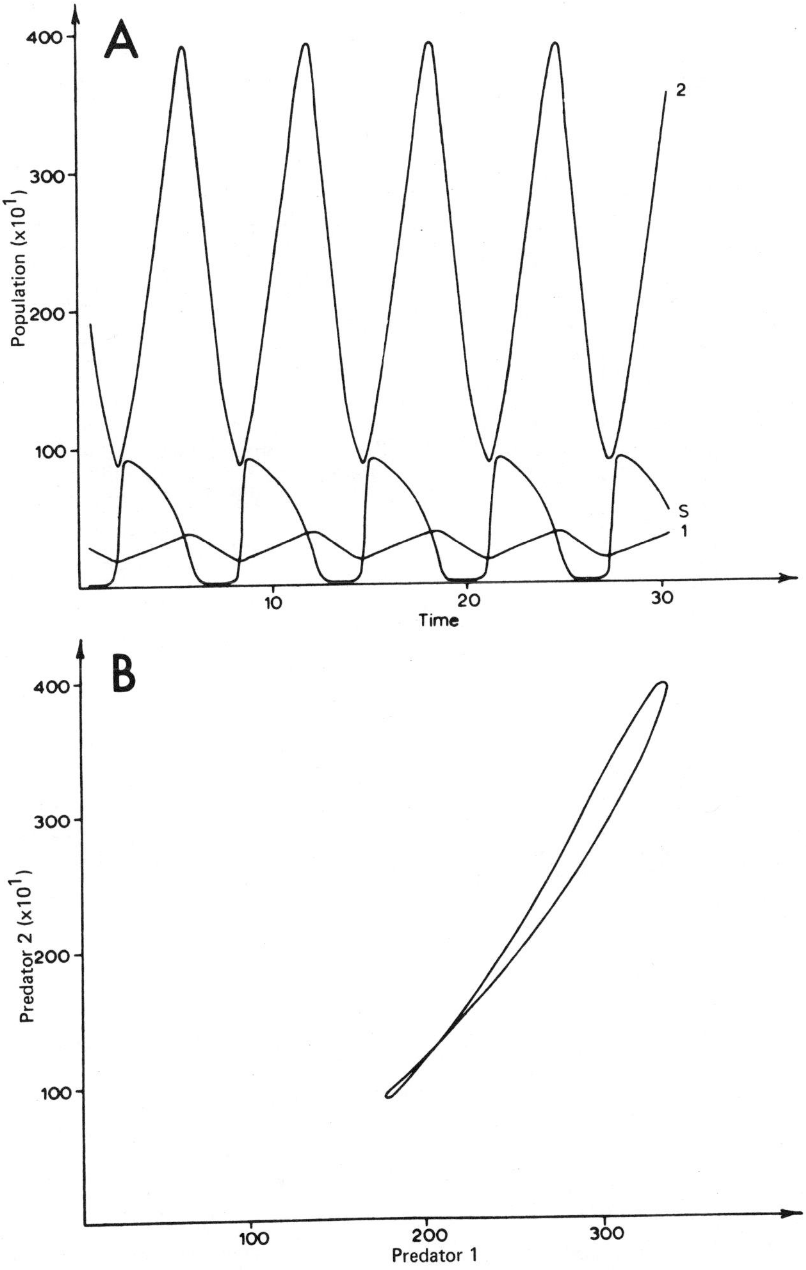

Fig. 4.6. Case in which there is oscillatory coesistence of 2 predator species (1 and 2) on a single prey species, S. Parameters are the same in Figure 1D with a_1 = 200 and K = 1100. A: Oscillations as in a function of time. B: Limit cycle of numbers of predator 1 plotted against numbers of predator 2. Initial values: x_1 = 307.13, x_2 = 2684.95, S = 8.60.

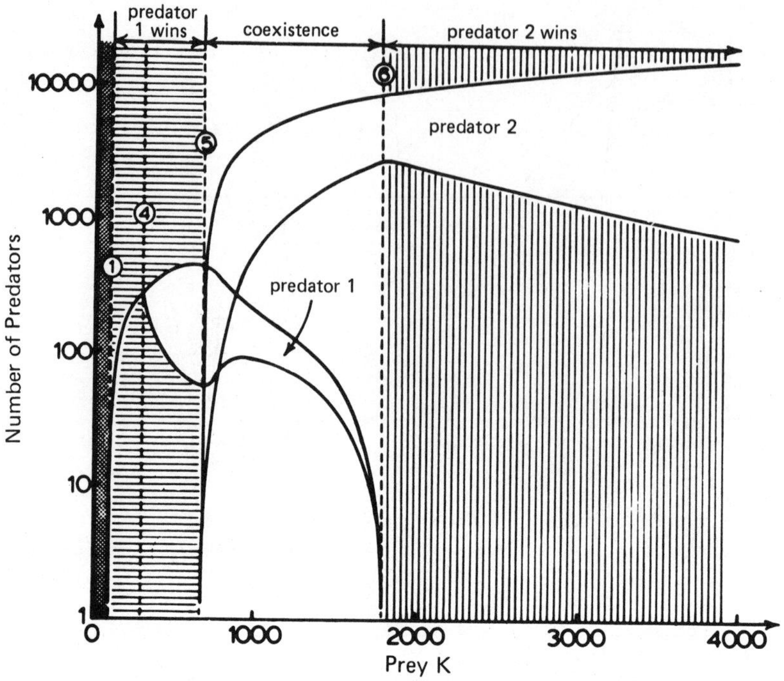

Fig. 4.7. Limiting behavior of 2 competing predators, one of which is a "K-strategist" (predator 1) and the other of which is an "r-strategist," preying upon a single prey population. Outcome as a function in Figure 4.5H with $a_1 = 100$. Shaded and hatched areas and numbered lines are codes as indicated in the legend and caption for Figure 4.5. Lines for predators 1 and 2 indicate the periodic maximal and minimal population sizes in the limiting oscillations.

the limit cycle in the open positive octant--which appears to be globally asymptotically stable--collapses into a limit cycle in the $S - x_2$ plane, retaining its global stability properties. This is the region where x_2 wins the competition. Proof that the two dimensional limit cycle bifurcates into a globally asymptotically stable limit cycle in the open octant is an open mathematical question and would appear to require information about the Poincaré map before existent bifurcation theorems could be applied. The collapse into the $S - x_2$ plane is the same problem viewed as K being sufficiently large but decreasing.

EXTENSION TO TWO RESOURCES

When the theory is extended to cover exploitative competition for two or more resources, it becomes necessary to consider how the resources, once consumed, interact to promote growth. Leon and Tumpson [30] and Hubbell and Hsu [21] have considered the cases of competition for two perfectly complementary or substitutable resources. The criteria for the outcomes for each case are given in [21]. We summarize these results in this section.

Before presenting the competition models for two species on two resources, it is necessary to discuss how the functional responses of the consumer species have been generalized from one to two resources. In the one-resource case, the per capita consumption rate, according to the Type II functional response, is given by $(m_{ri}/y_{ri})(R/a_{ri} + R))$ if the resource is R, or is given by $(m_{si}/y_{si})(S/a_{si} + S))$ if the resource is S. These one-resource per capita consumption rates can be rewritten as:

$$\frac{m_{ri}/a_{ri}}{y_{ri}}\,\frac{R}{1 + R/a_{ri}} \quad \text{and} \quad \frac{m_{si}/a_{si}}{y_{si}}\,\frac{S}{1 + S/a_{si}}$$

respectively. The generalization of two perfectly substitutable resources is well known, and corresponds in Michaelis-Menten theory to reaction rates with competitive inhibition--consumption of resource R acts as a competitive inhibitor in the consumption of resource S. In Holling's terminology, handling time devoted to processing a unit of resource R is time not available for the processing of resource S, and this competitive effect is linear. Therefore, the per capita consumption rates of resource R in the presence of substitutable resource S, and of S in the presence R, are given by:

$$\frac{m_{ri}/a_{ri}}{y_{ri}}\,\frac{R}{1 + R/a_{ri} + S/a_{si}} \quad \text{and} \quad \frac{m_{si}/a_{si}}{y_{si}}\,\frac{S}{1 + R/a_{ri} + S/a_{si}}$$

respectively. Note that the above expressions simplify to the previous case if one of the resources is absent.

The generalization of the functional response to two complementary resources is different. In this case, the per capita consumption rate of whichever resource is currently limiting growth is identical to the one-resource per capita consumption rate for the appropriate resource. The question then arises: At what rate is the nonlimiting resource consumed? This question can be answered by considering the yield of consumer produced per unit of resource consumed. When the yield factors, y_{ri} and y_{si} are constants, then it follows that there must be a fixed ratio of the growth-essential substances provided by resources R and S in a unit of consumer. Moreover, this also implies that the per capita consumption rate of the nonlimiting resource must be proportional to the per capita consumption rate of the limiting resource. If it were not, then the ratio of essential growth substances in the consumer would be changing, and the yield factors would no longer be constant. The proportionality constant is the ratio of the yield constants for the two resources. For example, suppose species i is S-limited. Then the per capita consumption rate of S, call it $f_1(S)$, is:

$$f_1(S) = \frac{m_{si}}{y_{si}} \frac{S}{a_{si} + S} \tag{11}$$

whereas the concurrent per capita consumption rate of the nonlimiting resource R is given by:

$$\left(\frac{y_{si}}{y_{ri}}\right) \cdot f_1(S) = \frac{m_{si}}{y_{ri}} \frac{S}{a_{si} + S} \tag{12}$$

Note that this expression does not contain the concentration of the nonlimiting resource R. Thus, it should be noted that: *For complementary resources R and S, when a species is S-limited, its per capita consumption rate of R is independent of the concentration of R; whereas, when the species is R-limited, its per capita consumption rate of S is independent of the concentration of S.*

For complementary resources, R and S, the species 1 and 2, competing exploitatively for them, the system of equations then is:

$$\frac{dS}{dt} = (S_0 - S)D_s - f_1(S, R)x_1 - f_3(S, R)x_x$$

$$\frac{dR}{dt} = (R_0 - R)D_r - f_2(S, R)x_1 - f_4(S, R)x_2$$

$$\frac{dx_1}{dt} = \left[\min\left\{\frac{m_{s_1}S}{a_{s_1} + S}, \frac{m_{r_1}R}{a_{r_1} + R}\right\} - D_1\right]x_1$$

$$\frac{dx_2}{dt} = \left[\min\left\{\frac{m_{s_2}S}{a_{s_2} + S}, \frac{m_{r_2}R}{a_{r_2} + R}\right\} - D_2\right]x_2$$

where S, R, N_1, and N_2 are all positive at time 0, and where:

$$f_1(S, R) = \begin{cases} \dfrac{m_{s_1}}{y_{s_1}} \dfrac{S}{a_{s_1} + S} & \text{if } \dfrac{m_{s_1}S}{a_{s_1} + S} \leq \dfrac{m_{r_1}R}{a_{r_1} + R} \\[2ex] \dfrac{m_{r_1}}{y_{s_1}} \dfrac{R}{a_{r_1} + R} & \text{if } \dfrac{m_{s_1}S}{a_{s_1} + S} \geq \dfrac{m_{r_1}R}{a_{r_1} + R} \end{cases}$$

$$f_2(S, R) = \begin{cases} \dfrac{m_{s_1}}{y_{r_1}} \dfrac{S}{a_{s_1} + S} & \text{if } \dfrac{m_{s_1}S}{a_{s_1} + S} \leq \dfrac{m_{r_1}R}{a_{r_1} + R} \\[2ex] \dfrac{m_{r_1}}{y_{r_1}} \dfrac{R}{a_{r_1} + R} & \text{if } \dfrac{m_{s_1}S}{a_{s_1} + S} \geq \dfrac{m_{r_1}R}{a_{r_1} + R} \end{cases}$$

$$f_3(S, R) = \begin{cases} \dfrac{m_{s_2}}{y_{s_2}} \dfrac{S}{a_{s_2} + S} & \text{if } \dfrac{m_{s_2}S}{a_{s_2} + S} \leq \dfrac{m_{r_2}R}{a_{r_2} + R} \\[2ex] \dfrac{m_{r_2}}{y_{s_2}} \dfrac{R}{a_{r_2} + R} & \text{if } \dfrac{m_{s_2}S}{a_{s_2} + S} \geq \dfrac{m_{r_2}R}{a_{r_2} + R} \end{cases}$$

$$f_4(S, R) = \begin{cases} \dfrac{m_{s_2}}{y_{r_2}} \dfrac{S}{a_{s_2} + S} & \text{if } \dfrac{m_{s_2} S}{a_{s_2} + S} \le \dfrac{m_{r_2} R}{a_{r_2} + R} \\ \dfrac{m_{r_2}}{y_{r_2}} \dfrac{R}{a_{r_2} + R} & \text{if } \dfrac{m_{s_2} S}{a_{s_2} + S} \ge \dfrac{m_{r_2} R}{a_{r_2} + R} \end{cases}$$

For substitutable resources, R and S, and species 1 and 2 competing exploitatively for them, the system of equations is:

$$\frac{dS}{dt} = (S_0 - S)D_s - g_1(S, R)x_1 - g_3(S, R)x_2$$

$$\frac{dR}{dt} = (R_0 - R)D_r - g_2(S, R)x_1 - g_4(S, R)x_2$$

$$\frac{dx_1}{dt} = \left[\frac{(m_{s_1}/a_{s_1})S + (m_{r_1}/a_{r_1})R}{1 + S/a_{s_1} + R/a_{r_1}} - D_1\right] \cdot x_1$$

$$\frac{dx_2}{dt} = \left[\frac{(m_{s_2}/a_{s_2})S + (m_{r_2}/a_{r_2})R}{1 + S/a_{s_2} + R/a_{r_2}} - D_2\right] \cdot x_2$$

where S, R, N_1, and N_2 are all positive at time 0, and where:

$$g_1(S, R) = \frac{m_{s_1}/a_{s_1}}{y_{s_1}} \frac{S}{1 + S/a_{s_1} + R/a_{r_1}}$$

$$g_2(S, R) = \frac{m_{r_1}/a_{r_1}}{y_{r_1}} \frac{R}{1 + S/a_{s_1} + R/a_{r_1}}$$

$$g_3(S, R) = \frac{m_{s_2}/a_{s_2}}{y_{s_2}} \frac{S}{1 + S/a_{s_2} + R/a_{r_2}}$$

$$g_4(S, R) = \frac{m_{r_2}/a_{r_2}}{y_{r_2}} \frac{R}{1 + S/a_{s_2} + R/a_{r_2}}$$

where,

S_0, R_0 = input concentrations of resource S and R, respectively,
D_s, D_r = input-output flow rate of medium containing S or R,
D_i = per capita death rate for the ith species,
m_{si}, m_{ri} = per capita birth rate of species i on resource S or R,
y_{si}, y_{ri} = yield of species i per unit of resource S on R consumed, and
a_{si}, a_{ri} = half-saturation constant for species i on resource S or R.

In the one-resource case, λ, the subsistence resource concentration for each competing species, was sufficient to predict the outcome of competition. The λ criterion is also important in the two-resource case, but in general it no longer provides sufficient information by itself to predict competitive outcomes, except in the case where one species has the lower λ's for both resources (this species wins regardless of initial abundance). In particular, additional competition criteria are required when species 1 has the lower subsistence concentration on one resource, but species 2 has the lower subsistence concentration on the second resource, because in this situation there are a number of possible outcomes. To discriminate between these additional outcomes, it becomes necessary to introduce two new parameters, T* and C_i, which differ depending on whether or not the resources are complementary or substitutable. The parameter C_i is species specific. These parameters for complementary resources are:

$$T^* = \frac{(R_0 - \ _{r_2})D_r}{(S_0 - \ _{s_1})D_s}, \qquad C_1 = \frac{1/y_{r_1}}{1/y_{s_1}}, \qquad C_2 = \frac{1/y_{r_2}}{1/y_{s_2}}$$

where λ_{r_i} and λ_{s_i} are the λ parameters of species i on resources r and s, respectively.

For substitutable resources, the corresponding values of T^*, C_1, and C_2 are:

$$C_1 = \frac{m_{r_1}/y_{r_1}}{m_{s_1}/y_{s_1}}, \qquad C_2 = \frac{m_{r_2}/y_{r_2}}{m_{s_2}/y_{s_2}}$$

$$T^* = \frac{\dfrac{(R_0 - R^*_{12})}{R^*_{12}} D_r}{\dfrac{(S_0 - S^*_{12})}{S^*_{12}} D_s}$$

where

$$R^* = \frac{\lambda_{r_1}\lambda_{r_2}(\lambda_{s_2} - \lambda_{s_1})}{\lambda_{r_1}\lambda_{s_2} - \lambda_{r_2}\lambda_{s_1}} \tag{13}$$

and

$$S^*_{12} = \frac{\lambda_{s_1}\lambda_{s_2}(\lambda_{r_1} - \lambda_{r_2})}{\lambda_{r_1}\lambda_{s_2} - \lambda_{r_2}\lambda_{s_1}}$$

The parameter T^* represents the ratio of the steady-state resource regeneration rates, R over S, when both species 1 and 2 are present at equilibrium. The C_1 and C_2 parameters represent the ratios of demand for resources R and S by species 1 and 2, respectively. In the case when one species has a lower λ for one of the resources, and the second species has the lower λ for the other resources, the outcome of competition depends on the relative rates of resource supply compared with the rates of resource demand by the two species. The 3-way inequalities involving T^*, C_1, and C_2 constitute criteria to resolve all remaining cases of exploitative competition for two resources. The T^* and C_i parameters differ in the cases of complementary and substitutable resources. In the complementary-resource case, the C_i's are ratios of the yield constants because the uptake rate of the nonlimiting resource is proportional to, and determined by, the uptake rate of the limiting resource. However, in the substitutable-resource case, the C_i's are ratios of per capita consumption rates instead. Since each substitutable resource is actually just an alternate source for the same essential nutrient, what becomes important to competitive outcomes in this case are the per capita rates of resource consumption by each species, in relation to the specific rates of resource regeneration. Relative rates of consumption are now important because neither species can be limited solely by one of the resources; thus, there is no simple refuge from competition by separate resource limitation in the two species.

When the present theory is extended from one resource to two, additional outcomes of competition not found in one-resource situations, are predicted. Two-resource theory predicts that there will be a broad region of parameter space for which the competing species can coexist, unlike the "knife-edge" condition of precisely equals λ's in one-resource theory. Moreover, two-resource theory predicts that, under certain conditions, the initial abundance of the competing species will determine which species is the eventual winner. Finally, the two-resource theory generates each of the classical outcomes of two-species competition in more than one way. These outcomes are compared in Table 4.2.

To our knowledge no one has yet reported a case of competition for two known complementary or substitutable resources in which the outcome depended on the initial abundances of the competitors. Tilman and Kilham [53] and Tilman [52] have performed interesting competition studies in semicontinuous cultures between two freshwater diatoms, *Asterionella formosa* Hass., and *Cyclotella meneghiniana* Kutz. for the complementary resources, phosphate and silicate. They did not report any cases in which the outcomes were dependent on initial numbers. However, they did find a broad region of coexistence over a range of ratios of silicate/phosphate in the influent supply to semicontinuous cultures of the two diatom species.

The data provided by Tilman [52] has been analyzed to see if there is any possibility of a case in which the initial number of *Asterionella* or *Cyclotella* could determine the outcome of competition. Let λ_{AP} and λ_{AS} be the λ criteria for *Asterionella* on phosphate and silicate, respectively; and let λ_{CP} and λ_{CS} be the λ criteria for *Cyclotella*, correspondingly. Assuming that all cell death was due to washout from the culture in the effluent, then the maximum death rate they studied experimentally was 0.5/day. If we use this rate, then the values of the λ criteria are: $\lambda_{AP} = 0.25\mu M$ (micromole), $\lambda_{AS} = 3.28\mu M$, $\lambda_{CP} = 0.417\mu M$, and $\lambda_{CS} = 0.90\mu M$. Thus, $\lambda_{AP} < \lambda_{CP}$, so that *Asterionella* has a lower subsistence concentration on phosphate than *Cyclotella* by more than an order of magnitude, but $\lambda_{CS} < \lambda_{AS}$, so that *Cyclotella* has a lower subsistence concentration on the silicate than *Asterionella*.

TABLE 4.2

Biological Classification of the Outcomes of Two-Resource Exploitative Competition Between Two Species

Biological Case	Lotka-Volterra Competition Criterion	Two-Resource Exploitative Competition Criteria
1. Species 1 always wins, regardless of initial density; species 2 dies out.	$\alpha < \frac{K_1}{K_2}$, $\beta > \frac{K_2}{K_1}$	(a) $\lambda_{s_1} < \lambda_{s_2}$, $\lambda_{r_1} < \lambda_{r_2}$ (b) $\lambda_{s_1} > \lambda_{s_2}$, $\lambda_{r_1} < \lambda_{r_2}$, $T^* < C_1, C_2$ (c) $\lambda_{s_1} < \lambda_{s_2}$, $\lambda_{r_1} > \lambda_{r_2}$, $T^* > C_1, C_2$
2. Species 2 always wins, regardless of initial density; species 1 dies out.	$\alpha > \frac{K_1}{K_2}$, $\beta < \frac{K_2}{K_1}$	(a) $\lambda_{s_1} > \lambda_{s_2}$, $\lambda_{r_1} > \lambda_{r_2}$ (b) $\lambda_{s_1} > \lambda_{s_2}$, $\lambda_{r_1} < \lambda_{r_2}$, $T^* > C_1, C_2$ (c) $\lambda_{s_1} < \lambda_{s_2}$, $\lambda_{r_1} > \lambda_{r_2}$, $T^* < C_1, C_2$
3. Species 1 and 2 persist in a stable coexistence.	$\alpha < \frac{K_1}{K_2}$, $\beta < \frac{K_2}{K_1}$	(a) $\lambda_{s_1} > \lambda_{s_2}$, $\lambda_{r_1} < \lambda_{r_2}$, $C_1 < T^* < C_2$ (b) $\lambda_{s_1} < \lambda_{s_2}$, $\lambda_{r_1} > \lambda_{r_2}$, $C_1 > T^* > C_2$
4. Species 1 wins, or species 2 wins, while rival species dies out; initial densities determine eventual winner.	$\alpha > \frac{K_1}{K_2}$, $\beta > \frac{K_2}{K_1}$	(a) $\lambda_{s_1} > \lambda_{s_2}$, $\lambda_{r_1} < \lambda_{r_2}$, $C_1 > T^* > C_2$ (b) $\lambda_{s_1} < \lambda_{s_2}$, $\lambda_{r_1} > \lambda_{r_2}$, $C_1 < T^* < C_2$

Next, it is necessary to compute T*, C_A, and C_C, where C_A and C_C are the criteria for *Asterionella* and *Cyclotella*, respectively, and

$$T^* = \frac{(P_0 - \lambda_{CP})D_P}{(S_0 - \lambda_{AS})D_S}$$

where P_0 and S_0 are the input phosphate and silicate concentrations, respectively D_P and D_S are the input flow rates of phosphate and silicate, which in this case are equal (taken as 0.5/day), and the point $(\lambda_{AS}, \lambda_{CP})$ is the intersection of the *Asterionella* and *Cyclotella* isoclines on the silicate-phosphate resource plane. Of the range of values of P_0 and S_0 tested by Tilman, we chose $P_0 = 10\mu M$ and $S_0 = 100\mu M$. This gives a value for T* = 0.100.

Finally, it is necessary to compute the C criteria for the two diatoms. The yield constants for *Asterionella* are reported by Tilman (1977) to be: $Y_{AP} = 2.18 \times 10^8$ cells/μM on phosphate, and $Y_{AS} = 2.51 \times 10^6$ cells/μM on silicate. Therefore, $C_A = (1/Y_{AP})/(1/Y_{AS}) = 1.15 \times 10^{-2}$. The yield constants for *Cyclotella* are: $Y_{CP} = 2.59 \times 10^7$ cells/μM on phosphate, and $Y_{CS} = 4.20 \times 10^6$ cells/μM on siiicate. Thus $C_C = (1/Y_{CP})/(1/Y_{CS}) = 1.62 \times 10^{-1}$.

With this information, the question of whether there can exist a case in which the winning diatom species (*Asterionella* or *Cyclotella*) is determined by the initial cell density of each diatom can be answered. Note that $\lambda_{AP} < \lambda_{CP}$ and $\lambda_{AS} > \lambda_{CS}$. Also note that $C_A < T^* < C_C$. This corresponds to a case of coexistence, a fact that Tilman [52] confirmed experimentally. In order for there to be a case in which the initial diatom density determines the outcome in this competitive system for these λ's, it would be necessary that the inequalities among C_A, T*, and C_C be totally reversed: $C_A > T^* > C_C$. This, in turn, would require substantial changes in the yield constants for phosphate and silicate in these two diatom species. Since only the criterion variable T* involves parameters under experimental control, there is no possibility of a case in which initial cell densities affect the competitive outcome between *Asterionella* and *Cyclotella*.

OTHER THEORETICAL WORK IN PROGRESS

A Four Population, Three Level Food Chain

The system (9) presumes a carrying capacity K but does not indicate the mechanism by which it occurs. One way in the laboratory to create the effect of a carrying capacity for the population S would be to grow it in a chemostat on a single resource, call it R. The equations for S and R without a predator for S would be given by $(5)_1$. Combining this with the ideas in the seventh section produces a three level, four population food chain whose equations are

$$R' = (R^{(0)} - R)D - \frac{m_3}{y_3}\frac{SR}{a_3 + R}$$

$$S' = \frac{m_3 SR}{a_3 + R} - DS - \sum_{i=1}^{2} \frac{m_i}{y_i}\frac{x_i S}{a_i + S} \tag{14}$$

$$x_i' = \frac{m_i x_i S}{a_i + S} - Dx_i, \qquad i = 1, 2$$

$$R(0) = R_0 > 0,\ S(0) = S_0 > 0,\ x_i(0) = x_{i0} > 0, \qquad i = 1, 2$$

Since death is through washout all of the death rates are the same, although one might want to allow individual death rates for mathematical completeness. The system (14) is currently being studied by the authors.

Delays

All of the above models assume instantaneous reaction, that is, there are no delays. The oscillations found in the experimental data in the sixth section suggest that delays may be present. In an experiment with the chemostat Caperon [4] was forced to consider delays in the chemostat equations in order to fit his experimental data. In fact, simple constant delays were not adequate and Caperon was forced to consider distributed delays. Currently the use of delays in $(5)_2$ is being investigated.

ACKNOWLEDGMENTS

The research of Stephen Hubbell was supported by National Science Foundation Grant DEB76-21409.

The research of P. Waltman was supported by National Science Foundation Grants MCS78-01097 and MCS79-01069.

These sources of support are gratefully acknowledged.

REFERENCES

1. J. R. Beckwith and D. Zipser (Eds.), *The Lactose Operon*. (Cold Spring Harbor Laboratories, Cold Spring Harbor, New York (1970).

2. G. Butler, "Coexistence in predator-prey systems," These Proceedings.

3. D. E. Burmaster, "The continuous culture of phytoplankton: mathematical equivalences among three steady state models," *Amer. Nat.* 113(1978), 123-134.

4. J. Caperon, "Time lag in population growth response of *Isochrysis galgana* to a variable nitrate environment," *Ecology* 50(1969), 188-192.

5. M. R. Droop, "The nutrient status of algal cells in continuous culture," *J. Marine Biol. Assoc. U. K.* 54(1974), 825-855.

6. R. W. Eppley and J. L. Coatsworth, "Uptake of nitrate and nitrite by *Pitylum brightwellii*--kinetics and mechanisms," *J. Phycol.* 4(1968), 151-158.

7. G. J. Gause, *The Struggle for Existence*. Williams and Wilkins, Baltimore, Maryland (1934).

8. D. E. Gill, "Intrinsic rates of increase, saturation densities, and competitive ability. I. An experiment with *Paramecium*," *Amer. Nat.* 106(1972), 461-471.

9. S. R. Hansen and S. P. Hubbell, "Single-nutrient microbial competition: agreement between experimental and theoretically forecast outcomes," Subm. *Science* (1979).

10. M. P. Hassell, *The Dynamics of Arthropod Predator-Prey Systems*. Princeton University, Princeton, New Jersey (1978).

11. M. P. Hassell, J. H. Lawton and J. R. Beddington, "Sigmoid functional responses by invertebrate predators and parasitoids," *J. Anim. Ecol.* 46(1977), 249-262.

12. D. Herbert, R. Elsworth and R. C. Telling, "The continuous culture of bacteria: a theoretical and experimental study," *J. Gen. Microbiol.* 14(1956), 601-622.

13. C. S. Holling, "The components of predation as revealed by a study of small mammal predation of the European pine sawfly," *Canad. Entomol.* 91(1959), 293-320.

14. C. S. Holling, "The functional response of predators to prey density and its role in mimicry and population regulation," *Mem. Entomol. Soc. Canada* 45(1965), 3-60.

15. C. S. Holling, "The functional response of invertebrate predators to prey density," *Mem. Entomol. Soc. Canada* 48(1966), 1-86.

16. S. B. Hsu, "Limiting behavior for competing species," *SIAM J. Appl. Math.* 34(1978), 760-763.

17. S. B. Hsu, "The application of the Poincaré-transform to the Lotka-Volterra model," *J. Math. Biology* 6(1978), 67-73.

18. S. B. Hsu, S. P. Hubbell and P. Waltman, "A mathematical theory of single-nutrient competition in continuous cultures of microorganisms," *SIAM J. Appl. Math.* 32(1977), 366-383.

19. S. B. Hsu, S. P. Hubbell and P. Waltman, "Competing predators," *SIAM J. Appl. Math.* 35(1978), 617-625.

20. S. B. Hsu, S. P. Hubbell and P. Waltman, "A contribution to the theory of competing predators," *Ecological Monographs* 48(1978), 337-349.

21. S. P. Hubbell and S. B. Hsu, "Exploitative competition for resources: multiple mechanistic origins for each outcome of classical competition theory," Subm. *Ecology*, 1979.

22. G. E. Hutchinson, *A Treatise on Limnology. II. Introduction to Lake Biology and Limnoplankton.* Wiley, New York (1967).

23. C. A. Istock, "Logistic interaction of natural populations of two species of water-boatmen," *Amer. Nat.* 111(1977), 279-287.

24. V. S. Ivlev, *Experimental Ecology of the Feeding of Fishes.* Yale University, New Haven, Connecticut (1961).

25. H. W. Jannash and R. T. Mateles, "Experimental bacterial ecology studied in continuous culture," *Advances in Microbial Physiology* 11(1974), 165-212.

26. J. Kalff and R. Knoechel, "Phytoplankton and their dynamics in oligotrophic and eutrophic lakes," *Ann. Rev. Ecol. Syst.* 9(1978), 475-495.

27. P. Kilham, "A hypothesis concerning silica and the freshwater planktonic diatoms," *Limnol. Oceanogr.* 16(1971), 10-18.

28. A. L. Koch, "Competitive coexistence of two predators utilizing the same prey under constant environmental conditions," *J. Theoret. Biol.* 44 (1974), 373-386.

29. J. P. LaSalle, *The Stability of Dynamical Systems*. Society for Industry and Appl. Math., Philadelphia (1976).

30. J. A. Leon and D. B. Tumpson, "Competition between two species for two complementary or two substitutable resources," *J. Theoret. Biol.* 50 (1975), 185-201.

31. H. R. Mahler and E. H. Cordes, *Biological Chemistry*. Harper and Row, New York (1966).

32. R. McGehee and R. A. Armstrong, "Some mathematical problems concerning the ecological principle of competitive exclusion," *J. Diff. Eq.* 23 (1977), 30-52.

33. R. S. Miller, "Pattern and process in competition," *Advances in Ecological Research* 4(1967), 1-74.

34. J. Monod, *Reserches sur la Croissance des Cultures Bacteriennes*. Hermann, Paris (1942).

35. J. Monod, "La technique de la culture continue: theorie et applications," *Ann. Inst. Pasteur* 79(1950), 390-410.

36. W. W. Murdoch, S. Avery and M. E. B. Smyth, "Switching in predatory fish," *Ecology* 56(1975), 1094-1105.

37. W. E. Neill, "The community matrix and interdependence of the competition coefficients," *Amer. Nat.* 108(1974), 399-408.

38. A. Novick and L. Szilard, "Description of the chemostat," *Science* 112 (1950), 215-216.

39. G. Oster and Y. Takahashi, "Models for age-specific interactions in a periodic environment," *Ecological Monographs* 44(1974), 483-501.

40. T. Park, "Beetles, competition, and populations," *Science* 138(1962), 1369-1375.

41. T. Park, P. H. Leslie and D. B. Mertz, "Genetic strains and competition in populations of *Tribolium*," *Physiol. Zoology* 37(1964), 97-162.

42. J. W. Payne, "Oligopeptide transport in *Escherichia coli*," *J. Biol. Chem.* 243(1968), 3395-3403.

43. L. A. Real, "The kinetics of functional response," *Amer. Nat.* 111(1977), 287-300.

44. R. C. Richmond, M. E. Gilpin, S. Perez Salas and F. J. Ayala, "A search for emergent competitive phenomena: the dynamics of multispecies *Drosophila* systems," *Ecology* 56(1975), 709-714.

45. T. Toyama, "Factors governing the hunting behavior and selection of food by the great tit (*Parus major* L.)," *J. Anim. Ecol.* 39(1970), 619-668.

46. G. W. Salt, "Predator and prey densities as controls of the rate of capture by the predator *Didinium nasutum*," *Ecology* 55(1974), 434-439.

47. T. W. Schoener, "Some methods for calculating competitive coefficients from resource utilization spectra," *Amer. Nat.* 108(1974), 320-340.

48. T. W. Schoener, "Alternatives to Lotka-Volterra competition models of intermediate complexity," *Theoret. Popul. Biol.* 10(1976), 309-333.

49. N. W. Taylor, "A mathematical model for two *Tribolium* populations in competition," *Ecology* 49(1968), 843-848.

50. P. A. Taylor and J. L. Williams, "Theoretical studies on the coexistence of competing species under continuous-flow conditions," *Canad. J. Microbiol.* 21(1975), 90-98.

51. D. W. Tempest, "The place of continuous culture in microbiological research," *Advances in Microbial Physiol.* 4(1970), 223-245.

52. D. Tilman, "Resource competition between planktonic algae: an experimental and theoretical approach," *Ecology* 58(1977), 338-348.

53. D. Tilman and S. S. Kolham, "Phosphate and silicate growth and uptake kinetics of the diatoms *Asterionella formosa* and *Cyclotella meneghiniana*," *J. Phycol.* 12(1976), 375-383.

54. L. Tinbergen, "The natural control of insects in pinewoods. I. Factors influencing the intensity of predation by songbirds," *Arch. Neerl. Zool.* 13(1960), 266-336.

55. D. Titman, "Ecological competition between algae: experimental confirmation of resource-based competition theory," *Science* 192(1977), 463-465.

56. P. van den Ende, "Predator-prey interactions in continuous culture," *Science* 181(1973), 562-564.

57. J. H. Vandermeer, "The competitive structure of communities: an experimental approach with protozoa," *Ecology* 50(1969), 362-371.

58. H. Veldcamp, "Ecological studies with the chemostat," *Advances in Microbiol Ecology* 1(1977), 59-95.

59. P. F. Verhulst, "Notice sur la loi que la population pursuit dans son accroissement," *Correspond. Math. Phys.* 10(1938), 113-121.

60. V. Volterra, "Variations and fluctuations of the number of individuals of animal species living together," in *Animal Ecology*, R. N. Chapman (Ed.), McGraw-Hill, New York (1926).

61. P. J. Wangersky, "Lotka-Volterra population models," *Ann. Rev. Ecol. Syst.* 9(1978), 189-218.

62. H. Wilbur, "Competition, predation and the structure of the *Ambystoma-Rana sylvatica* communitie," *Ecology* 53(1972), 3-21.

63. T. Yoshida, "Studies on the interspecific competition between bean weevils," *Mem. Fac. Liberal Arts Educ. Akita Univ. Natur. Sci.* 20(1966), 59-98.

A LIAPUNOV FUNCTIONAL FOR A CLASS OF REACTION-DIFFUSION SYSTEMS

Nicholas D. Alikakos

Department of Mathematics
Purdue University
West Lafayette, Indiana

INTRODUCTION

In this chapter we study certain aspects of the asymptotic behavior of the solutions of the system

$$\frac{\partial u_i}{\partial t} = \nu_i \, \Delta u_i + B_i(u_1,\ldots, u_N) \qquad \text{on } \Omega x(0, \infty) \tag{1}$$

$$\left.\frac{\partial u_i}{\partial n}\right|_{\partial\Omega} = 0 \qquad \text{on } \partial\Omega x(0, \infty) \tag{2}$$

$$u_i(x, 0) = f(x) \tag{3}$$

where Ω is a bounded smooth connected set in $\mathbb{R}^n$, Δ is the Laplacian in $\mathbb{R}^n$, $\partial/\partial n$ is the Neuman operator and ν_i are positive constants.

We would like now to digress and recall some standard results for the equation

$$\frac{\partial u}{\partial t} = \Delta u + f(x, u) \tag{4}$$

together with Neuman or Dirichlet boundary conditions for the sake of an analogy which will become clear in what follows.

It is well known that (4) has a natural Liapunov Functional [6] given by

$$V(g) = \frac{1}{2}\int_{\Omega}|\nabla g|^2 dx - \int_{\Omega} F(x, g(x))dx, \; F(x, z) = \int_0^z f(x, s)ds \tag{5}$$

By means of this functional and the Invariance Principle [7, 5] one can assert that under appropriate conditions on f every solution of (4) stabilizes, that is, approaches a solution $\hat{u}$ of the associated elliptic problem

$$\Delta\hat{u} + f(x, \hat{u}) = 0 \tag{6}$$

with corresponding boundary conditions. Thus the functional (5) can be visualized as a means of comparing solutions of (4) with those of (6).

It turns out that in certain cases one can construct a Liapunov Functional for (1), (2), (3) which in effect "compares" solutions of this system with those of the associated system of ordinary differential equations

$$\frac{d\phi_i}{dt} = B_i(\phi_1, \ldots, \phi_N) \qquad i = 1, \ldots, N \tag{7}$$

Our objective is to study various aspects of this comparison.

Specifically in the first section we introduce the relevant Liapunov Functional motivated from Williams, Chow [10]* and then we employ the Invariance Principle for abstract semigroups and state the main result.

*A. Hastings [12] has independently obtained a class of Liapunov functionals which are more general than those of Williams and Chow [10], but less general than the ones in this work.

In the second section we obtain, as applications, results that generalize those obtained by Williams, Chow [10], and Leung [11]. The flavor of the results here is that (1), (2), (3) and (7) have identical ω limit sets independently of the diffusion coefficients.

In the third section we investigate the type of information that can be obtained for (1), (2), (3) once an invariant region for (7) is known. In contrast to the second section the results here depend heavily on the diffusion coefficients and more precisely on the relationship of the diffusion matrix with the geometry of the region. Connections with the theory of invariant rectangles [3] are discussed.

For an exposition of some of the concepts used here we refer the reader to Dafermos [4] where a wealth of further references is given.

A LEMMA

The concept of the semigroup is central in our considerations and for this purpose we need to recall certain related facts for the system (1), (2), (3). It is well known [6] that this can be written abstractly as an ordinary differential equation

$$\frac{du}{dt} + Au = B(u) \tag{8}$$

where

$$A = -\begin{pmatrix} \nu_1\Delta & \cdots & 0 \\ \vdots & & \\ 0 & \cdots & \nu_N\Delta \end{pmatrix}, \quad B(u) = \begin{pmatrix} B_1(u_1, \ldots, u_N) \\ \vdots \\ B_N(u_1, \ldots, u_N) \end{pmatrix}$$

on the Banach space $X = L^P = L^P(\Omega)x\ldots xL^P(\Omega)$. In this setting A is a closed densely defined operator that generates an analytic semigroup and with domain

$$\prod_{i=1}^{N} D(-\Delta), \quad D(-\Delta) = \left\{ \xi \in W^{2,P}(\Omega) / \frac{\partial \xi}{\partial n}\bigg|_{\partial\Omega} = 0 \right\}$$

For measuring regularity one uses the fractional spaces X^α, where

$$X^\alpha = \prod_{i=1}^{N} X_i^\alpha, \quad X_i = L^p(\Omega)$$

and α is a number in $[0, 1]$. In this scale of smoothness $X_i^1 = D(-\Delta)$ equipped with the graph norm of $-\Delta + I$ and $X_i^0 = L^p(\Omega)$. It can be shown [6] that $X^\alpha \hookrightarrow C^\nu(\Omega) x \ldots x C^\nu(\Omega)$ for α such that

$$2\alpha - \frac{n}{p} \geq \nu > 0, \qquad \alpha < 1 \tag{9}$$

Under the hypotheses of local Lipschitz continuity on the B_i's one can appeal to the abstract theory of semilinear parabolic equations in Banach spaces [6] and secure the existence and uniqueness of a local solution. Having introduced now the setting we can proceed to the matter of interest.

Consider the system of ordinary differential equation associated with (1), (2), (3) and given by

$$\frac{d\phi_i}{dt} = B_i(\phi_1, \ldots, \phi_N), \qquad i = 1, \ldots, N \tag{10}$$

and let us assume that (10) has a continuous Liapunov Function $E(z_1, \ldots, z_N)$. The following set of hypotheses is useful.

(H.1) E is twice continuously differentiable and $H(z) = \left(\frac{\partial^2 E(z)}{\partial z_i \partial z_j}\right)$ is positive semidefinite.

(H.2) The symmetrization of $\begin{pmatrix} \nu_1 & & & \\ & \nu_2 & & \\ & & \ddots & \\ & & & \nu_N \end{pmatrix} H(z)$ is positive definite except, possibly, at finitely many points.

(H.3) The symmetrization of $\begin{pmatrix} \nu_1 & & \\ & \ddots & \\ & & \nu_N \end{pmatrix} H(z)$ is positive semidefinite.

Now we can state

LEMMA 1 Define the functional

$$V(f_1(\cdot),\dots, f_N(\cdot)) = \int_\Omega E(f_1(x),\dots, f_N(x))dx \tag{11}$$

$f(\cdot) = (f_1(\cdot),\dots, f_N(\cdot))$ in X^α, α as in (9). Then under the hypothesis (H.1), (H.3) V is a Liapunov Functional for (1), (2), (3). Moreover, given a relatively compact orbit $\{u(\cdot, t)|t > 0\}$ (in X^α) we have that $u(\cdot, t)\xrightarrow[X^\alpha]{} M$, where M is the maximal invariant set of (1), (2), (3) that consists of functions which lie in X^α and take values in $E^* = \{(z_1,\dots, z_N)|\dot{E}(z) = 0\}$. If (H.3) is replaced by the stronger (H.2) then we can assert that $u(\cdot, t)\xrightarrow[X^\alpha]{} M^*$ where M^* is the maximal invariant set of (10) in E^*.

REMARKS

1. By $\dot{E}(z)$ we mean the grad $E \cdot B(z)$.

2. Note that if (H.1), (H.2) hold, then any relatively compact orbit approaches the maximal invariant set of the system of ordinary differential equations and that the convergence, by (9), is quite strong. It turns out that the solutions of the system (1), (2), (3) in the case where $B_i(z_1,\dots, z_N) = z_iC_i(z_1,\dots, z_N)$ and C_i satisfy a "food pyramid condition" (a concept introduced by Williams, Chow [10]) are always global and if in addition an L^1, uniform in time, a priori bound is known for them then it can be guaranteed that the corresponding orbits are relatively compact in any X^β, $\beta \in [0, 1)$. Moreover they become positive if the initial data are nonnegative. These facts will be of use to us later on. For proofs of these statements we refer the reader to [1].

3. It turns out that the condition of positivity on the symmetrization of

$$\begin{pmatrix} \nu_1 & & \\ & \ddots & \\ & & \nu_N \end{pmatrix} H(z)$$

is essential and that in general (H.1) is not sufficient by itself to secure that the solutions of (1), (2), (3) approach those of the associated (10). By elementary facts from linear algebra ([13], p. 427) this positivity as a rule, imposes restrictions on the ratios ν_i/ν_j, k, j = 1,..., N.

Sketch of Proof of Lemma 1 A calculation yields

$$\frac{dV}{dt} = \sum_{i=1}^{N} \int \frac{\partial E}{\partial z_i} B_i(u_1,\ldots, u_N)dx - \int \mathrm{trace}\{(\nabla u)^T \begin{pmatrix} \nu_1 & & \\ & \ddots & \\ & & \nu_N \end{pmatrix} H(\nabla u)\}dx \quad (12)$$

The first term on the right of (12) is nonpositive by virtue of E being a Liapunov Function for (10). The integrand of the second term is nonnegative by (H2) or (H3). The rest of the argument follows from the Invariance Principle. We refer the reader to [1] for details. It should be noted that the above computation can be carried equally well for the case when

$$\begin{pmatrix} \nu_1 & & \\ & \ddots & \\ & & \nu_N \end{pmatrix}$$

is replaced by an arbitrary constant positive definite matrix D.

APPLICATIONS

Consider the classical Lotka-Volterra Predator-Prey model given by the system of ordinary differential equations

$$\begin{aligned} \frac{d\phi_1}{dt} &= (\alpha - \phi_2)\phi_1 \\ \frac{d\phi_2}{dt} &= (\phi_1 - \beta)\phi_2 \end{aligned} \quad (13)$$

with α and β positive constants.

It is well known that all solutions of (13) are periodic and that there is a first integral given by

$$E(x, y) = [\alpha\beta]^{-1}\{[x - \beta - \beta \ell n(\beta^{-1}x)] + [y - \alpha - \alpha \ell n(\alpha^{-1}y)]\}$$
$$= E_1(x) + E_2(y) \tag{14}$$

Note that $E_i''(z) > 0$ for $z > 0$.

Consider the associated diffusion system

$$\frac{\partial u_1}{\partial t} = \nu_1 \Delta u_1 + (\alpha - u_2)u_1$$
$$\frac{\partial u_2}{\partial t} = \nu_2 \Delta u_2 + (u_1 - \beta)u_2 \tag{15}$$

$$\left.\frac{\partial u_i}{\partial n}\right|_{\partial\Omega} = 0 \tag{16}$$

with initial conditions

$$f_i(x) \geq 0 \tag{17}$$

Williams, Chow [10] showed that any solution of (15), (16), (17) that happens to be uniformly bounded for all time approaches, as $t \to \infty$, a periodic solution of (13). According to Remark 2 in the first section this assumption of boundedness is redundant since an L^1 bound is available. Since the Hessian $\partial^2 E/\partial z_i \partial z_j$ in the case at hand is diagonal (H.2) follows trivially from (H.1) and thus by Lemma 1 we conclude that (15), (16), (17) and (13) have identical ω-limit sets. Taking advantage of the specific Liapunov Function we can show more as the following theorem shows.

THEOREM 1* Any solution of the system (15), (16), (17) with nonnegative nontrivial conditions in X^α (i.e., sufficiently smooth), α as in (9), approaches exponentially in the X^β sense (for any $\beta \in [0, 1]$) a periodic solution of (13) with asymptotic phase. Moreover, the manifold of periodic solutions is orbitally stable with asymptotic phase and amplitude.

*This result is not entirely expected in view of the fact that (18) is not structurally stable.

For the proof we refer the reader to [1].

In [11] Leung considers the following system

$$\frac{\partial u_1}{\partial t} = \nu_1 \Delta u_1 + u_1(a - bu_1 - cu_2), \qquad \text{on } \Omega x(0, \infty) \tag{18}$$

$$\frac{\partial u_2}{\partial t} = \nu_2 \Delta u_2 + u_2(pu_1 - qu_2 - r), \qquad \text{on } \Omega x(0, \infty) \tag{19}$$

$$\left.\frac{\partial u_i}{\partial n}\right|_{\partial\Omega} = 0$$

$$u_i(x, 0) = f_i(x) \geq 0$$

where all the constants are assumed positive and moreover $ap > br$. He shows, using methods in the spirit of Williams, Chow [10] that every solution with nonnegative initial data converges uniformly to (u^*, v^*), where

$$u_1^* = \frac{aq + cr}{cp + bq}, \qquad u_2^* = \frac{ap - br}{cp + bq}$$

To obtain this result we note, following Leung, that

$$E(x, y) = \left[px - pu_1^* \ln\left(\frac{x}{pu_1^*}\right)\right] + \left[cy - cu_2^* \ln\left(\frac{y}{cu_2^*}\right)\right]$$

is a Liapunov Function for the associated system

$$\begin{aligned} \frac{d\phi_1}{dt} &= \phi_1(a - b\phi - c\phi_2) \\ \frac{d\phi_2}{dt} &= \phi_2(p\phi_1 - q\phi_2 - r) \end{aligned} \tag{20}$$

Clearly E is convex with diagonal Hessian. Thus hypotheses (H1), (H2) of Lemma 1 are automatically satisfied (one, as in the previous example, has to consider the case of positive functions in X as the phase space). Since (u_1^*, u_2^*) is the unique point in E* it follows that all solutions of (18), (19) converge at least in the Hölder continuous sense to (u^*, v^*).

It should be noted that (18), (19) is of the type mentioned in Remark 2 in the first section and thus all orbits are relatively compact. Positivity of the solutions also follows by the same remark.

SIGNIFICANCE OF (H3)

In the applications considered so far the Hessian of E has been diagonal and thus (H2) or (H3) are trivially satisfied once (H1) is. The problem of finding invariant sets for (1), (2), (3) once existence of such sets has been established for (7) presents itself as an example where (H3) plays a crucial role. More precisely consider the set (or more generally a finite intersection of such sets)

$$M^c = \{z \in \mathbb{R}^N / G(z) \leq c\}$$

where G is a C^2 function from $\mathbb{R}^N$ to $\mathbb{R}$, nonnegative, and c is a number and assume that M^c is positively invariant or globally attracting with respect to the flow that (7) induces. A natural question to ask is whether this set retains its invariance properties with respect to (1), (2), (3). In their fundamental paper Chueh, Conley, Smoller [3] showed that in general the the only sets M^c that behave as such are the rectangles with sides parallel to the axes. In the next few pages we investigate certain more quantitative analogs of this statement and more precisely we establish simple relationships between the diffusion coefficients and the geometry of the set M^c so that invariance, in some form, persists. To fix ideas we need some definitions and terminology. First assume for convenience that $m(\Omega) = 1$. We will say that the set M^c is *invariant in the mean* (I.I.M) *of order* p with respect to (1), (2), (3) if the set

$$M_p^c = \left\{ \xi(\cdot) \in [C(\Omega)]^N \;\middle|\; \left(\int_\Omega G^p(\xi(x))\,dx \right)^{1/p} \leq c \right\}$$

is positively invariant. The set M^c is said I.I.M. if p = 1 in the preceding statement. The next proposition establishes the necessity of (H3) for this type of invariance to hold.

PROPOSITION 1 Let $c \in (\text{Inf } G, \sup G)$. If the set M^c is I.I.M. then by necessity the symmetrization of

$$\begin{pmatrix} \nu_1 & & & \\ & \nu_2 & & \\ & & \ddots & \\ & & & \nu_N \end{pmatrix} \left(\frac{\partial^2 G(z)}{\partial z_i \partial z_j}\right)$$

is positive semidefinite for every z and moreover $\nabla G \cdot B \leq 0$ on ∂M^c.

We refer for the proof to [2].

REMARKS

1. In the case of the heat equation

$$\frac{\partial u}{\partial t} = \Delta u$$

$$\left.\frac{\partial u}{\partial n}\right|_{\partial\Omega} = 0$$

it can be easily seen that M^c is I.I.M. only if G is convex. In fact let $\xi_0(\cdot)$ on ∂M^c and consider the corresponding solution u(x, t). Since M^c is positively invariant it follows that $\lim_{t\to\infty} u(x, t)$ (which always exists) should be in M^c. However

$$\lim_{t\to\infty} u(x, t) = \int_\Omega \xi_0(x)\,dx \qquad (m(\Omega) = 1)$$

and thus we should have

$$G\left(\int \xi_0(x)\,dx\right) \leq c = \int G(\xi_0(x)\,dx$$

which of course is to be expected for every ξ_0 only if G is convex.

2. The condition of positivity on the symmetrization of

$$\begin{pmatrix} \nu_1 & & \\ & \ddots & \\ & & \nu_N \end{pmatrix} \left(\frac{\partial^2 G}{\partial z_i \partial z_j} \right)$$

implies by a fact from linear algebra ([13], p. 427) the following restriction on the ratio of the diffusion coefficients.

$$\sqrt{\frac{\nu_i}{\nu_j}} + \sqrt{\frac{\nu_j}{\nu_i}} \leq 2 \frac{\sqrt{\dfrac{\partial^2 G}{\partial z_i^2} \dfrac{\partial^2 G}{\partial z_j^2}}}{\left| \dfrac{\partial^2 G}{\partial z_i \partial z_j} \right|}$$

The following corollary establishes that the only sets M^c that can be I.I.M. are the convex ones.

COROLLARY 1 If the set M^c is I.I.M. for some c in (Inf G, sup G) then by necessity G is a convex function.

Proof By the previous proposition we have that the symmetrization of

$$\begin{pmatrix} \nu_1 & & \\ & \ddots & \\ & & \nu_N \end{pmatrix} \left(\frac{\partial^2 G(z)}{\partial z_i \partial z_j} \right)$$

for every z has to be positive semidefinite. Let z be fixed and take

$$B = \begin{pmatrix} \nu_1 & & \\ & \ddots & \\ & & \nu_N \end{pmatrix}, \quad A = \left(\frac{\partial^2 G(z)}{\partial z_i \partial z_j} \right)$$

Thus AB + BA = C where C is positive semidefinite. Clearly the corollary will be established if we show that A is positive semi-definite. For this purpose consider the ordinary differential equation $x' = (-A)x$. Then the function $V(x) = \langle Bx, x \rangle$, as it can be easily checked, is a Liapunov Function for this equation. By well known elementary facts from the theory of stability we obtain that Re $\sigma(-A) \leq 0$ and since A is symmetric the result follows.

It is interesting that convexity enters from different considerations in [3].

The following proposition establishes the special nature of strips that are parallel to a coordinate axis in $\mathbb{R}^N$ (and thus of the rectangles with sides parallel to coordinate axes) as well as the special nature of equal diffusion coefficients. This result is the analog in the present setting of Theorem 1.4 in [3].

PROPOSITION 2

1. If M^c is invariant in the mean for every p then it is positively invariant.

2. Let M^c be invariant in the mean of order p for every p then by necessity if $\nu_i \neq \nu_j$ one of the partials $\partial G/\partial z_i$, $\partial G/\partial z_j$ has to vanish identically. In particular if the diffusion coefficients are distinct then G can depend at most on one coordinate of z.

Proof

1. This is immediate since if

$$\left(\int G^p(u(x, t))dx\right)^{\frac{1}{p}} \leq c$$

for every p it follows that $\max_x G(u(x, t)) \leq c$.

2. Fix p, t and apply Proposition 1 wich c replaced by c^p and G(z) replaced by $G_p(z) = G^p(z)$. Then we obtain that the symmetrization of

$$\begin{pmatrix} \nu_1 & & \\ & \ddots & \\ & & \nu_N \end{pmatrix} \left(\frac{\partial^2 G_p(z)}{\partial z_i \partial z_j}\right)$$

is positive semi-definite (≥ 0) for every z. Computing the symmetrization of this matrix we obtain that

$$p(p - 1)G^{p-2}\left(\frac{\nu_i + \nu_j}{2} \frac{\partial G}{\partial z_j} \frac{\partial G}{\partial z_i}\right) + p\ G^{p-1}\left(\frac{\nu_i + \nu_j}{2} \frac{\partial^2 G}{\partial z_i \partial z_j}\right) \geq 0$$

From this by simplifying we have that

$$(p - 1)\left(\frac{\nu_i + \nu_j}{2} \frac{\partial G}{\partial z_i} \frac{\partial G}{\partial z_j}\right) + \left(\frac{\nu_i + \nu_j}{2} \frac{\partial^2 G}{\partial z_i \partial z_j}\right) G \geq 0$$

This inequality has to hold for every z and for every p. Thus by necessity the first matrix has to be positive semidefinite. By a standard fact from linear algebra then we have that for every z

$$\nu_i \left(\frac{\partial G}{\partial z_i}\right)^2 \nu_j \left(\frac{\partial G}{\partial z_j}\right)^2 \geq \left(\frac{\nu_i + \nu_j}{2}\right)^2 \left(\frac{\partial G}{\partial z_i}\right)^2 \left(\frac{\partial G}{\partial z_j}\right)^2$$

which of course is true only if $\nu_i = \nu_j$ or in the case $\nu_i \neq \nu_j$ if $\partial G/\partial z_i$ or $\partial G/\partial z_j$ vanish identically.

APPLICATION Next we illustrate the applicability of invariance in the mean by an example that is related to the theory of superconductivity of liquids. We refer the reader to the example D in [3] where a special case with equal diffusion coefficients is treated. Let A be a positive definite symmetric matrix and define

$$G(z) = \langle Az, z \rangle$$

Consider the system

$$\frac{\partial u_i}{\partial t} = \nu_i \Delta u_i + (c - G(u))u_i \qquad \text{on } \Omega x(0, \infty) \tag{21}$$

$$\left.\frac{\partial u_i}{\partial n}\right| = 0 \qquad \text{on} \qquad \partial\Omega \times (0, \infty) \qquad i = 1, \ldots, N$$

$(m(\Omega) = 1)$. Let the diffusion coefficients satisfy

$$\left(\frac{\max_k \nu_k}{\min_k \nu_k}\right)^{\frac{1}{2}} \leq \frac{\sqrt{E_A} + 1}{\sqrt{E_A} - 1} \qquad \text{(H)} \tag{22}$$

where E_A = spectral condition number of A $\overset{df}{=}$ $(\max_k \lambda_k(A))/(\min_k \lambda_k(A))$.

Then under hypothesis (H)

1. each ellipsoid

$$M^k = \{z/G(z) \leq k\} \qquad (k \geq c)$$

is invariant in the mean for (21). Moreover all orbits are relatively compact and all the ω-limit sets are in $M^c = \{\xi(\cdot)/\int G(\xi(x))dx \leq c\}$.

2. If all the diffusion coefficients are equal then the sets M^k are positively invariant and the critical ellipsoid M^c is approached by the solutions exponentially in the L^∞ sense.

COMMENT Part 1 can be obtained as an easy application of the theory in [3]. We include it here for illustrating our point of view.

Proof For the first part we have

$$\frac{d}{dt}\int_\Omega G(u(x, t))dx$$

$$= -\int_\Omega \mathrm{tr}\left\{(\nabla u)^T \begin{pmatrix} \nu_1 & & \\ & \ddots & \\ & & \nu_N \end{pmatrix} \frac{\partial^2 G}{\partial z_i \partial z_j} (\nabla u)\right\} dx + \sum_{i=1}^N \int \frac{\partial G}{\partial z_i} B_i \, dx$$

$$= -\int_\Omega \mathrm{tr}\left\{(\nabla u)^T \begin{pmatrix} \nu_1 & & \\ & \ddots & \\ & & \nu_N \end{pmatrix} A(\nabla u)\right\} dx + 2\int_\Omega (c - G(u))G(u)dx \tag{23}$$

At this stage we need the following result from linear algebra.

LEMMA 2 (Nicholson [8]) Let A, B be two positive Hermitian matrices and define C by C = AB + BA. Then C is positive definite if

$$E_B < \frac{(\sqrt{E_A} + 1)^2}{(\sqrt{E_A} - 1)^2}$$

where E_Q = spectral condition number of matrix Q.

Applying this lemma we note that under hypothesis (H) the symmetrization of

$$\begin{pmatrix} \nu_1 & & \\ & \ddots & \\ & & \nu_N \end{pmatrix}$$

A is positive semidefinite and thus obtain that

$$\frac{d}{dt}\int_\Omega G(u(x, t))dx \le 2\int_\Omega (c - G(u))G(u)dx$$

$$\le 2[c \int G(u)dx - (\int G(u)dx)^2] \tag{24}$$

At this point we obtain the first part, for if $\xi \in \partial M^k$ and $u(x, 0) = \xi(x)$, then

$$\left.\frac{d}{dt}\right|_{t=0} \int G(u(x, t))dx \le 2[ck - k^2] \le 0$$

Solving (24) (and assuming for the moment global existence)

$$\overline{\lim_{t\to\infty}} \int G(u(x, t))dx \le c \tag{25}$$

To obtain global existence as well as compactness of the orbits note first that $\int G(u)dx$ is bounded. Thus $\int |u|^2dx$ is bounded. So by applying Theorem 3.1 in [1] to each equation in (21) we obtain an *a priori* L^∞ bound for u. By known regularity arguments [6] global existence and compactness in any X^{α_0} $(0 \le \alpha_0 < 1)$ follows and thus part 1 is established by virtue of (25).

To prove the second part we let $G_p(z) = (G(z))^p$. Then

$$\frac{d}{dt}\int G_p(u)dx$$

$$= -\int tr\left\{(\nabla u)^T \begin{pmatrix} \nu_1 & & \\ & \ddots & \\ & & \nu_N \end{pmatrix} \left(\frac{\partial^2 G_p}{\partial z_i \partial z_j}\right) (\nabla u)\right\}dx + \sum_{i=1}^N \int \frac{\partial G_p}{\partial z_i} B_i\, dx \le 0$$

Since the first term is nonpositive we have

$$\frac{d}{dt}\int G^p(u)dx \le 2p\int G^p(u)(c - G(u))dx$$

$$= 2p[c\int G^p(u)dx - \int G^{p+1}(u)dx]$$

$$= 2p[c\int G^p(u)dx - \int (G^p(u))^{\frac{p+1}{p}} dx]$$

$$\le 2p[c\int G^p(u)dx - (\int G^p(u)dx)^{\frac{p+1}{p}}] \tag{26}$$

At this point we obtain that all M^k are I.I.M. of order p for every p. For let $\xi \in M_p^k$. Then by (26) for u(x, t) with u(x, 0) = ξ(x) we have

$$\left.\frac{d}{dt}\right|_{t=0}\int G^p(u)dx \le 2p[ck^p - k^{p+1}] = 2pk^p[c - k] \le 0$$

Thus, we obtain that each M^k is invariant. Finally, let us solve the differential inequality (26). After some calculations we obtain that

$$\int G^p(u(x, t))dx^{\frac{1}{p}} \le \frac{c}{1 + \left[\dfrac{c - (\int G_p(u(x, 0))dx)^{\frac{1}{p}}}{(\int G_p(u(x, 0))dx)^{\frac{1}{p}}}\right]e^{-ct}}$$

Taking the limit as $p \to \infty$

$$\max_{\Omega} G(u(x, t)) \le \frac{c}{1 + \left[\dfrac{c - \max G(u(x, 0))}{\max G(u(x, 0))}\right]e^{-ct}}$$

REMARKS

1. It is rather interesting that the flatter the ellipsoid is the more restrictive condition (22) becomes and conversely as the ellipsoid tends to a ball the restriction disappears.

2. It turns out that in most applications where a convex set $C \subseteq \mathbb{R}^N$ attracts orbits of (7) the geometrical condition

$$B(z) \cdot (z - z_0) < 0 \qquad z \notin C$$

is satisfied, where z_0 is the point on ∂C that realizes the distance of z from C. It is not hard to show [2] that this implies the square of the distance function from the set C is a Liapunov Function for (7). In this context then the distinguishing feature of the strips parallel to a coordinate axis is easy to see: simply for these figures (and only for these) the corresponding Liapunov Function has diagonal Hessian and thus (H3) imposes no restrictions on the diffusion coefficients.

REFERENCES

1. N. D. Alikakos, "An application of the invariance principle to Reaction-Diffusion equations," *J. D. E.* 39(1979).

2. N. D. Alikakos, "Remarks on invariance in Reaction-Diffusion equations," (to appear).

3. K. N. Chueh, C. C. Conley and J. A. Smoller, "Positively invariant regions for systems of Diffusion Equations," *I. U. M. J.* 26(1977).

4. C. M. Dafermos, "Asymptotic behavior of solutions of evolution equations, in *Nonlinear Evolution Equations*, M. Crandall (Ed.), Academic Press (1979).

5. J. K. Hale, "Dynamical systems and stability," *J. Math. Anal. Appl.* 26(1969).

6. D. Henry, *Geometric theory of parabolic equations* (monograph to appear).

7. J. P. LaSalle, "Stability theory and the asymptotic behavior of dynamical systems," in *Dynamic stability and structures. Proc. International Conference*, Pergamon, Long Island City, New York (1966).

8. D. W. Nicholson, "Eigenvalue bounds for AB + BA, with A, B positive definite matrices," *Linear Algebra and Appl.* 24(1979).

9. H. Weinberger, "Invariant sets for weakly coupled parabolic and elliptic systems," *Rend. Mat. Univ. Roma.* 8(1975).

10. S. Willialms and P. L. Chow, "Nonlinear Reaction-Diffusion models," *J. Math. Anal. Appl.* 62(1978).

11. A. Leung, "Limiting behavior for a prey-predator model with diffusion and crowing effects," *J. M. B.* 6(1978).

12. A. Hastings, "Global stability in Lotka-Volterra systems with diffusion," *J. M. B.* (1978).

13. B. Noble and J. W. Daniel, *Applied Linear Algebra*, 2nd edition, Prentice Hall, New Jersey (1977).

STOCHASTIC PREY-PREDATOR RELATIONSHIPS

Georges A. Bécus

Engineering Science Department
University of Cincinnati
Cincinnati, Ohio

INTRODUCTION

Half a century ago, Lotka [24] and Volterra [40] initiated independently the deterministic theory of population dynamics. In the following decennium Kostitzin [20] and Kolmogorov [19] furthered the study of the nonlinear Lotka-Volterra model to the point that little has been added to the deterministic theory in the ensuing forty years. For an excellent account of the deterministic theory the reader should consult [31].

The stochastic theory of population dynamics was initiated by Feller almost forty years ago. Since then a vast amount of works dealing with stochastic models for population studies has appeared. The justification for such stochastic models can be found in the books by Bartlett [3], Iosifescu and Tautu [17], May [25], and Goel and Richter-Dyn [13].

This article presents a survey of various stochastic models (and their analysis) of interacting populations in a prey-predator relationship. These various works have been regrouped under six general headings. (A seventh one regroups hard-to-classify works.) However the reader will soon discover that the demarcation line between these six groups is not as clearly defined as the table of contents would indicate.

Caveat lector! This survey is restricted to prey-predator interactions. A number of works dealing with competitive interactions could be adapted to the prey-predator interaction. I have not attempted to do so. As any survey, this one undoubtedly suffers from omissions for which I apologize. I have tried to keep the same notation throughout (thus X_1 and X_2 denote sizes of the prey and predator populations respectively) while at the same time I endeavored to stay as close as possible to the notation of the original works to facilitate the reader consultation of these sources. I have assumed the reader has some familiarity both with the deterministic theory of population dynamics and with applied probability and stochastic processes.

BIRTH AND DEATH PROCESS MODELING

The earliest stochastic model of interacting populations in a prey-predator relationship is probably due to Chiang [11] and is based on birth and death processes. (See also [9], Section 4.2D, and [14], Section 7, for an account of these results.)

More precisely, let $X_1(t)$ and $X_2(t)$ denote the (random) sizes of the prey (S_1) and predator (S_2) populations respectively, and let $P(x_1, x_2; t) = P\{X_1(t) = x_1, X_2(t) = x_2\}$, $x_i = 0, 1, \ldots$, $i = 1, 2$, be their joint probability distribution. Let us assume that

(i) the probability of a unit increase in the size X_i of population S_i in the time interval $(t, t + \Delta t)$, given that there are x_i individuals in S_i and x_j individuals in S_j at time t, is $\lambda_i(x_i, x_j)\Delta t + 0(\Delta t)$, $i = 1, 2$, $i \neq j = 1, 2$;

(ii) the probability of a unit decrease in the size X_i of population S_i in the time interval $(t, t + \Delta t)$, given that there are x_i individuals in S_i and x_j individuals in S_j at time t, is $\mu_i(x_i, x_j)\Delta t + 0(\Delta t)$, $i = 1, 2$, $i \neq j = 1, 2$;

(iii) the probability of a change (increase or decrease) of absolute value greater than one in $(t, t + \Delta t)$ is $0(\Delta t)$ for S_1 and S_2.

Then $\{X_1(t), X_2(t)\}$ is a bivariate birth and death process the joint distribution of which satisfies the differential equation

$$\frac{d\,P(x_1, x_2; t)}{dt}$$

$$= -[\lambda_1(x_1, x_2) + \mu_1(x_1, x_2) + \lambda_2(x_1, x_2) + \mu_2(x_1, x_2)]P(x_1, x_2; t)$$

$$+ \lambda_1(x_1 - 1, x_2) \quad P(x_1 - 1, x_2; t) + \mu_1(x_1 + 1, x_2)P(x_1 + 1, x_2; t)$$

$$+ \lambda_2(x_1, x_2 - 1)P(x_1, x_2 - 1; t) + \mu_2(x_1, x_2 + 1)P(x_1, x_2 + 1; t)$$

$$x_i \geq 0, \qquad i = 1, 2 \tag{1}$$

and the obvious condition

$$P(x_1, x_2; t) = 0 \quad \text{if} \quad x_1 \quad \text{and/or} \quad x_2 < 0$$

Let us assume, in analogy with the deterministic case that

$$\lambda_1(x_1) = \lambda_1\, x_1, \quad \mu_1(x_1, x_2) = \mu_1\, x_1\, x_2$$

$$\lambda_2(x_1, x_2) = \lambda_2\, x_1\, x_2, \quad \mu_2(x_2) = \mu_2\, x_2 \tag{2}$$

where λ_1, λ_2, μ_1, μ_2 are nonnegative constants. Substitution of (2) into (1) and introduction of the generating function

$$F(r_1, r_2; t) = \sum_{x_1, x_2 = 0}^{\infty} P(x_1, x_2; t) r_1^{x_1} r_2^{x_2} \tag{3}$$

leads to the partial differential equation

$$\frac{\partial F}{\partial t} = \lambda_1\, r_1(r_1 - 1) \frac{\partial F}{\partial r_1} - \mu_2(r_2 - 1) \frac{\partial F}{\partial r_2}$$

$$+ [\lambda_2 r_1 r_2 (r_2 - 1) + \mu_1 r_2 (1 - r_1)] \frac{\partial^2 F}{\partial r_2 \partial r_1} \tag{4}$$

Unfortunately, the general solution of equation (4) is difficult to obtain.

Nonetheless, from it one can obtain the following hierarchy of differential equations for the moments of X_1 and X_2

$$\frac{dE\{X_1\}}{dt} = \lambda_1 E\{X_1\} - \mu_1 E\{X_1 X_2\} \tag{5a}$$

$$\frac{dE\{X_2\}}{dt} = \lambda_2 E\{X_1 X_2\} - \mu_2 E\{X_2\} \tag{5b}$$

$$\frac{dE\{X_1 X_2\}}{dt} = (\lambda_1 + \lambda_2 - \mu_1 - \mu_2)E\{X_1 X_2\} + \lambda_2 E\{X_1^2 X_2\} - \mu_1 E\{X_1 X_2^2\} \tag{6}$$

etc.

Upon comparison of system (5) with the deterministic Lotka-Volterra system

$$\begin{aligned} \frac{dx_1}{dt} &= \lambda_1\, x_1 - \mu_2\, x_1\, x_2 \\ \frac{dx_2}{dt} &= \lambda_2\, x_1\, x_2 - \mu_2\, x_2 \end{aligned} \tag{7}$$

one sees that as $E\{X_1 X_2\} \neq E\{X_1\}E\{X_2\}$ the expected solutions to the stochastic model will be different from the solution to the deterministic one. Furthermore, equations (5) could have been obtained directly from (7) by replacing the deterministic population sizes x_i by the random ones X_i and taking expectation. This approach, amounting to a randomization of the deterministic model, does not appear to have been followed by anybody to date.

The above hierarchy of equations for the moments of X_1 and X_2 can be solved approximately using truncated hierarchy techniques (see [36], Section 8.6, and [32] and the references therein). For example, upon setting $E\{X_1 X_2\} = E\{X_1\}E\{X_2\}$ system (5) becomes equivalent to system (7); a higher order approximation could be obtained upon setting $E\{X_1^2 X_2\} = E\{X_1\}E\{X_1 X_2\}$ and $E\{X_1 X_2^2\} = E\{X_1 X_2\}E\{X_2\}$ in which case (5-6) reduces to a system of three nonlinearly coupled equations.

Alternatively, Monte Carlo simulations of the birth and death process could be used [1, 2, 21, 22, 23].

For other more recent results on birth and death and related processes applicable to prey-predator interactions the reader may consult [17], Section 3.3, [28], [30] and [4] which are discussed in more details in the fourth section and [37].

FOKKER-PLANCK EQUATION METHODS

Another approach consists in assuming that the random fluctuations affecting the prey-predator relationship are of "white noise" type. In this case the population sizes are diffusion processes whose (joint) probability distribution is the solution to a Fokker-Planck (or Kolmogorov, or diffusion) equation.

For example in [14] (cf. also [13], Section 6.2 for an account of these results) the equation for the size X_i of population S_i is taken to be of the form

$$\frac{dX_i}{dt} = k_i X_i G(X_i/\theta_i) + X_i\{\beta_i^{-1} \sum_{\substack{j=1 \\ j\neq i}}^{n} a_{ij} X_j + U_i(t)\} \tag{8}$$

$$i = 1,\ldots, n$$

where k_i, θ_i, and β_i^{-1} are the growth rate, saturation level, and equivalence number* respectively for S_i while the a_{ij}'s are interaction constants satisfying $a_{ij} = -a_{ji}$. The function G is the saturation inducing term taken of the form

$$G(x/\theta) = [1 - (x/\theta)^{\alpha}]/\alpha \tag{9}$$

so that for $\alpha = 1$ we obtain the Verhulst interaction $1 - x/\theta$ and for $\alpha = 0$ we obtain the Gompertz interaction $\ell n(x/\theta)$ (cf. [14, 13] for more details). Finally $U_i(t)$ accounts for random influences.

*The term "equivalence number" was introduced by Volterra. During interaction of populations S_i and S_j, the ratio of increase (decrease) in X_i to increase (decrease) in X_j per unit time is $\beta_i^{-1}/\beta_j^{-1}$.

Since the population sizes are random, the entire term in brackets in (8) may be considered as a random function $F_i(t)$ influencing X_i so that (8) can be rewritten as (dropping the subscript i as the equations for each population will all be of the same form)

$$\frac{dX}{dt} = k\ X\ G(X/\theta) + X\ F(t) \tag{10}$$

Assuming F to be of "white noise" type with $E\{F\} = 0$ and $E\{F(t)F(t+\tau)\} = \sigma^2\delta(\tau)$, the Fokker-Planck equation for the transition probability density $P(x, t|x_o)$ of the process satisfying (10) is easily found to be

$$\frac{\partial P}{\partial t} = -\frac{\partial}{\partial x}[a(x)P] + \frac{1}{2}\frac{\partial^2}{\partial x^2}[b(x)P] \tag{11}$$

where

$$a(x) = k\ x\ [1 - (x/\theta)^\alpha]/\alpha + \sigma^2\ x/2$$

$$b(x) = \sigma^2\ x^2$$

The steady state ($t = \infty$) solution of (11) was found to be, [14],

$$P(x) = \alpha[\nu(x/\theta)^\alpha]^\nu \exp[-\nu(x/\theta)^\alpha]/x\Gamma(\nu) \tag{12}$$

where $\nu = 2\ k/\sigma^2\alpha^2$ which for $\alpha = 0$ (Gompertz interaction) reduces to

$$P(x) = (k/\pi\sigma^2)^{1/2}\ x^{-1}\ \exp\{-k[\ell n(x/\theta)]^2/\sigma^2\}$$

and for $\alpha = 1$ (Verhulst interaction) to

$$P(x) = [\Gamma(2k/\sigma^2)x]^{-1}(2kx/\theta\sigma^2)^{2k/\sigma^2}\ \exp(-2kx/\theta\sigma^2)$$

where Γ is the gamma function.

Using statistical mechanical arguments the time dependent distribution was found to be in the Gompertz case $\alpha = 0$, [14].

$$P(x, t|x_o) = \left\{\frac{\theta}{\pi\sigma^2 x^2(1 - e^{-2kt})}\right\}^{1/2} \exp\left\{-\frac{k[\ln u]^2}{\sigma^2(1 - e^{-2kt})}\right\} \tag{13}$$

where

$$u = \frac{x}{\theta}\left(\frac{\theta}{x_o}\right)e^{-kt}$$

In the general case ($\alpha \neq 0$) approximate expressions were obtained in three regimes: (i) population far from saturation, (ii) population fluctuating around the expected steady state value, and (iii) regimes other than (i) and (ii). From these expressions as well as from (12) and (13) the moments of X can be evaluated (see [13, 14] for details).

In [25] the prey-predator equations are taken of the form (for an n-species system)

$$\frac{dX_i}{dt} = F(X_1(t), X_2(t),\ldots, X_n(t)), \qquad i = 1,\ldots, n$$

where the F_i's incorporate some white noise environmental fluctuations. The joint probability distribution $f(x_1, x_2,\ldots, x_n; t)$ satisfies the Fokker-Planck equation

$$\frac{\partial f}{\partial t} = -\sum_{i=1}^{n} \frac{\partial}{\partial x_i}(M_i f) + \frac{1}{2}\sum_{i,j=1}^{n} \frac{\partial^2}{\partial x_i \partial x_j}(V_{ij} f)$$

where

$$M_i = E\{F_i\}$$

and

$$V_{ij} = E\{(F_i - M_i)(F_j - M_j)\}$$

In particular, it is noted that the results of [27] are applicable to a one-predator-one-prey system with Lotka-Volterra interactions, environmental stochasticity and density-dependent term in the prey growth rate.

In [26] May also uses the Fokker-Planck equation approach to derive a stability criterion. He takes the prey predator system

$$\frac{dX_i}{dt} = F_i[X_1(t),\dots, X_n(t); \kappa_1(t),\dots, \kappa_m(t)], \qquad i = 1,\dots, n \tag{14}$$

where the interaction parameters κ_j, $j = 1,\dots, m$ are of the form

$$\kappa_j(t) = \kappa_j^{o} + \gamma_j(t)$$

with γ_j a zero mean white noise and $\mathrm{Cov}(\gamma_i \gamma_j) = \sigma_{ij}^2$. Writing $X_i(t) = X_i^*[1 + Y_i(t)]$ where X_i^* is the solution of

$$0 = F_i[X_1^*,\dots, X_n^*; \kappa_1^o,\dots, \kappa_m^o], \qquad i = 1,\dots, n$$

i.e. is the equilibrium in the absence of random fluctuations, and expanding (14) about X_i^* leads, upon keeping linear terms, to

$$\frac{dY_i}{dt} = \sum_j a_{ij}\, Y_j(t) + \sum_j \mu_{ij}\, \gamma_j(t)$$

where the interaction matrix A is defined by

$$a_{ij} = \frac{X^*_j}{X^*_i}\left(\frac{\partial F_i}{\partial X_j}\right)_{X_k = X^*_k}$$

The Fokker-Planck equation takes the form

$$\frac{\partial f}{\partial t} = -\sum_{i,j} \frac{\partial}{\partial y_i}(a_{ij}\, y_j\, f) + \frac{1}{2}\sum_{i,j} d_{ij} \frac{\partial^2 f}{\partial y_i \partial y_j}$$

from which May derived in the case when A is symmetric and $\sigma_{ij} = \sigma\delta_{ij}$ the stability criterion

$$[-\lambda]_{min} > \sigma^2 \tag{15}$$

where $[-\lambda]_{min}$ is the negative real part of the eigenvalue of A nearest the imaginary axis. Criterion (15) says that provided the random fluctuations (as measured by σ^2) are not too large, the stochastic system (14) is stable in the sense that population paths will not wander away too far from the equilibrium configuration $(X_1^*, \ldots, X_n^*)$. (Cf. The eighth section for more on the stability criterion (15).)

THE REACTED PROCESS APPROACH

Several authors have considered the special case of prey-predator interaction in which one population influences the growth of the other without being itself influenced by the other.* We briefly review these works.

Becker [4] used birth, death and immigration processes to model a stochastic analogue of

$$\frac{dx_1}{dt} = \nu_1 + (\lambda_1 - \mu_1)x_1 - \alpha\, x_1\, x_2$$

$$\frac{dx_2}{dt} = \nu_2 + (\lambda_2 - \mu_2)x_2$$

He found that the joint probability distribution satisfies the Kolmogorov forward equation

$$\frac{dP(x_1, x_2; t)}{dt}$$

$$= [\nu_1 + \lambda_1(x_1 - 1)]P(x_1 - 1, x_2; t) + [\nu_2 + \lambda_2(x_2 - 1)P(x_1, x_2 - 1); t)$$

$$- (\nu_1 + \nu_2 + \lambda_1\, x_1 + \lambda_2\, x_2 + \mu_1\, x_1 + \mu_2\, x_2 + \alpha\, x_1\, x_2)P(x_1, x_2; t)$$

$$+ (\mu_1 + \alpha\, x_2)(x_1 + 1)P(x_1 + 1, x_2; t) + \mu_2(x_2 + 1)P(x_1, x_2 + 1; t)$$

*The process describing the evolution of the influenced population was termed "reacted process" by Kannan [18].

from which it is found that the factorial generating function

$$F_n(t, z) = \sum_{x_1, x_2=0}^{\infty} x_1^{(n)} z^{x_2} P(x_1, x_2; t) \tag{16}$$

where $x_1^{(n)} = x_1(x_1 - 1)\ldots(x_1 - n + 1)$ and $|z| \leq 1$, satisfies the differential difference equation

$$\frac{\partial F_n}{\partial t} - [\lambda_2 z^2 - (\alpha n + \lambda_2 + \mu_2] \frac{\partial F_n}{\partial z}$$

$$= [\nu_2 z + n\lambda_1 - n\mu_1 - \nu_2]F_n + n[(n - 1)\lambda_1 + \nu_1]F_{n-1} \tag{17}$$

and the boundary condition

$$F_n(0, z) = x_{10}^{(n)} z^{x_{20}}$$

where $x_{i0} = X_i(0)$, $i = 1, 2$

The use of the factorial generating function (16) as opposed to the usual generating function (3) leads to considerable simplification.

Becker obtained a recursive expression for the solution to (17) in the case when $\lambda_2 = 0$, i.e., when there are no births in population S_2, and computed the first few moments of the distribution both for the time independent and time homogeneous cases. In particular he studied the limits as $t \to \infty$ of $E\{X_1\}$, $E\{X_2\}$ and $Cov(X_1, X_2)$. He also considered a model with the additional interaction term $\beta x_1 x_2$ in the growth equation of S_1.

Puri [30] considered the same model as Becker [4] including the added interaction $\beta x_1 x_2$ but with $\nu_1 \equiv \nu_2 \equiv 0$. Since the process $\{X_2(t), t \geq 0\}$ influences the growth of the process $\{X_1(t), t \geq 0\}$ without being itself influenced by this latter process, he based his approach on the conditional study of X_1 given a realization of X_2, then took expectation over the process X_2 to uncondition X_1.

He evaluated the generating function (3) of the bivariate process assuming all the rates involved (λ_i, μ_i, α, β) are constant and in the special case for which $\lambda_1/\beta = \mu_1/\alpha = \theta$. The cases $\alpha < \beta$, $\alpha = \beta$, and $\alpha > \beta$ were treated separately. The calculations involved integrals of the birth and death process X_2. The moments of the process as well as the distributions of the time to extinction and of the duration of interaction between X_1 and X_2 were also studied.

Kannan [18] studied the evolution of the predator population S_2 under the influence of a free prey population S_1. He considered $X_1(t)$ to be a birth and death process while $X_2(t)$ is governed by

$$\frac{dX_2(t)}{dt} = -a\, X_2(t) + b\, X_1(t)\, X_2(t), \qquad t > 0$$

$$X_2(0) = X_{20}$$

The reacted predator population process $\{X_1(t), X_2(t), t \geq 0\}$ is shown to be a strong Markov process whose classification into transient/recurrent processes follows from that of the prey birth and death process. A necessary and sufficient condition for the existence of a steady state distribution for the reacted predator population process is given without particular reference to the prey population. Furthermore it is shown that if the prey birth and death process has a steady-state distribution then the reacted predator population process has a steady state distribution which satisfies a second-order difference equation. Finally it is shown that if $a/b = a^*$, a positive integer, then whenever $X_1(t) = a^*$, $d\, X_2(t)/dt = 0$, that is the reacted predator process takes rest from its evolution until the next interaction. If in particular a^* is a recurrent state of the prey process, then the predator process returns to infinitely many rests so that the asymptotic saturating probability for the predator population is $[\lambda_{a^*}\, m(a^*)]^{-1}$ where λ_{a^*} is the birth rate for $X_1(t)$ given $X_1(0) = a^*$ and $m(a^*)$ is the expected return time to a^* of $X_1(t)$.

ITÔ STOCHASTIC EQUATION MODELS

When the random fluctuations affecting the prey-predator relationship are of "white noise" type we can make use of the Fokker-Planck equation methods reviewed in the third section. Alternatively, realizing that the equations in this case (cf. Eq. (10) for example) are Itô stochastic differential equations one may use the vast body of results concerning these stochastic differential equations (see for example [8], Chap. 7, and the references therein) to study the prey-predator equations.

Gard and Kannan [12] used Itô equations techniques to study the following random version of the Lotka-Volterra equations

$$\begin{aligned} dX_1 &= b_1(X_1, X_2)dt + \sigma_1(X_1, X_2)dW_1(t) \\ dX_2 &= b_2(X_1, X_2)dt + \sigma_2(X_1, X_2)dW_2(t) \end{aligned} \tag{18}$$

where

$$b_1(X_1, X_2) = X_1(b_{11} - b_{12}X_2 - b_{13}X_1)$$

$$b_2(X_1, X_2) = X_2(-b_{21} + b_{22}X_1 - b_{23}X_2)$$

with $b_{ij} \geq 0$, $i = 1, 2$, $j = 1, 2, 3$, for a prey-predator system, σ_i, $i = 1, 2$, are scalar functions, and W_i, $i = 1, 2$, are independent standard Brownian motions.

They study the asymptotic behavior ($t \to \infty$) of (18) or its vector form

$$d\underset{\sim}{\xi}(t) = \underset{\sim}{b}(\underset{\sim}{\xi}(t))dt + \underset{\sim}{\sigma}(\underset{\sim}{\xi}(t))d\underset{\sim}{\beta}(t)$$

where

$$\underset{\sim}{\xi} = \begin{bmatrix} X_1 \\ X_2 \end{bmatrix}, \quad \underset{\sim}{b} = \begin{bmatrix} b_1 \\ b_2 \end{bmatrix}, \quad \underset{\sim}{\sigma} = \begin{bmatrix} \sigma_1 & 0 \\ 0 & \sigma_2 \end{bmatrix}, \quad d\underset{\sim}{\beta} = \begin{bmatrix} dW_1 \\ dW_2 \end{bmatrix}$$

subject to the initial condition $\underset{\sim}{\xi}(0) = \underset{\sim}{\xi}_o$, and in a sufficiently large but bounded domain D containing the equilibrium or saturation level solution $\underset{\sim}{\xi}^*$. Thus $\underset{\sim}{\xi}^*$ satisfies $\underset{\sim}{b}(\underset{\sim}{\xi}^*) = \underset{\sim}{0}$ and $\underset{\sim}{\sigma}(\underset{\sim}{\xi}^*) = \underset{\sim}{0}$.

Let $K \subset D$ be a compact set with $\underset{\sim}{\xi}_o$, $\underset{\sim}{\xi}^* \in K$ then (i) almost all population paths which do not exit from K in any finite time approach the equilibrium level $\underset{\sim}{\xi}^*$ as $t \to \infty$, and (ii) almost all population paths which do exit from K in a finite time reach the boundary of D as $t \to \infty$ leading to extinction or explosion (beyond a certain level) of one or both populations. The above results are contained in Theorems 3.1 and 3.2 of [12]. Note in particular that almost no population path is cyclic.

In addition, the probabilities of saturation, extinction, and explosion conditioned on the initial population level $\underset{\sim}{\xi}_o$ are solutions of the Dirichlet problem in a disk $\bar{B}$ with $B \subset D - \{\underset{\sim}{\xi}^*\}$ for the elliptic operator

$$L \equiv \sum_{i=1}^{2} \{a_i(x_1, x_2) \frac{\partial^2}{\partial x_i^2} + b_i(x_1, x_2) \frac{\partial}{\partial x_i} \}$$

associated to (18).

For a general discussion on stochastic differential equations of Itô and Stratonovich types as models in population biology the reader is referred to [38].

THE RANDOM EQUATION APPROACH

All the approaches discussed so far assume either a special kind of stochastic processes (birth and death process, white noise process) and/or stochastic models which are noisy versions of the deterministic one (third and fifth sections). In the random equation approach there is no a priori restriction on the stochastic processes and the stochastic model is a randomized version of the deterministic one in which the various rate constants are assumed random.

Prajneshu [29] considered a random version of the Gompertzian model of prey predator interaction

$$\frac{dX_1}{dt} = \alpha_1 X_1 - \beta_1 X_1 \ln X_2 - \gamma_1 X_1 \ln X_1$$

$$\frac{dX_2}{dt} = -\alpha_2 X_2 + \beta_2 X_2 \ln X_1 - \gamma_2 X_2 \ln X_2$$

which can be linearized upon introduction of the new dependent variables $Y_i = \ln X_i$, $i = 1, 2$, to give

$$\begin{cases} \dfrac{dY_1}{dt} = \alpha_1 - \beta_1 Y_2 - \gamma_1 Y_1 \\ \\ \dfrac{dY_2}{dt} = -\alpha_2 + \beta_2 Y_1 - \gamma_2 Y_2 \end{cases} \tag{19}$$

The various rate constants are assumed to be nonnegative random processes.

Assuming one source of randomness affects all interaction rates, these can be written as

$$\beta_i(t) = \beta_{i0}[1 + \varepsilon_i \Delta(t)], \qquad i = 1, 2$$

$$\gamma_i(t) = \gamma_{i0}[1 + \varepsilon_{i+2} \Delta(t)], \qquad i = 1, 2$$

where $\beta_{i0} = E\{\beta_i\}$, $\gamma_{i0} = E\{\gamma_i\}$, $i = 1, 2$, and $\Delta(t)$ is a zero mean second order stochastic process. System (19) can then be written in vector form as

$$\frac{d\underset{\sim}{Y}}{dt} = \underset{\sim}{A}\,\underset{\sim}{Y}(t) + \Delta(t)\,\underset{\sim}{B}\,\underset{\sim}{Y}(t) + \underset{\sim}{R}(t) \tag{20}$$

where

$$\underset{\sim}{Y}(t) = \begin{bmatrix} Y_1(t) \\ Y_2(t) \end{bmatrix}, \quad \underset{\sim}{A} = \begin{bmatrix} -\gamma_{10} & -\beta_{10} \\ \beta_{20} & -\gamma_{20} \end{bmatrix},$$

$$\underset{\sim}{B} = \begin{bmatrix} -\varepsilon_3\gamma_{10} & -\varepsilon_1\beta_{10} \\ \varepsilon_2\beta_{20} & -\varepsilon_4\gamma_{20} \end{bmatrix}, \quad \underset{\sim}{R}(t) = \begin{bmatrix} \alpha_1(t) \\ \alpha_2(t) \end{bmatrix}$$

and the solution of which is

$$\underset{\sim}{Y}(t) = e^{\underset{\sim}{A}t}\,\underset{\sim}{Y}(0) + \int_0^t e^{\underset{\sim}{A}(t-t')}\,\Delta(t')\,\underset{\sim}{B}\,\underset{\sim}{Y}(t')dt'$$

$$+ \int_0^t e^{\underset{\sim}{A}(t-t')}\,\underset{\sim}{R}(t')dt' \tag{21}$$

Equation (21) can be solved by the method of successive approximations. The first iteration leads upon taking expectation to the following equation for E{Y}

$$E\{\underset{\sim}{Y}(t)\} = e^{\underset{\sim}{A}t}\, \underset{\sim}{Y}(0) + \int_0^t e^{\underset{\sim}{A}(t-t')} \cdot \underset{\sim}{B}\left[\int_0^{t'} e^{\underset{\sim}{A}(t'-t'')}\, \underset{\sim}{B}\, E\{\Delta(t')\,\Delta(t'')\,Y(t'')\}dt'' + \int_0^{t'} e^{\underset{\sim}{A}(t'-t'')}\, E\{\Delta(t')\underset{\sim}{R}(t'')\}dt''\right] + \int_0^t e^{\underset{\sim}{A}(t-t')}\, E\{\underset{\sim}{R}(t')\}dt' \tag{22}$$

which unfortunately contains the unknown third order correlation functions $E\{\Delta(t')\Delta(t'')Y(t'')\}$.

Prajneshu resolved this problem known as the closure problem by assuming Δ and R to be uncorrelated and taking Δ to be a dichotomic Markov process in which case (22) reduces to

$$E\{\underset{\sim}{Y}(t)\} = e^{\underset{\sim}{A}t}\, \underset{\sim}{Y}(0) + \int_0^t e^{\underset{\sim}{A}(t-t')}\, \underset{\sim}{B}\, \underset{\sim}{\phi}(t')dt' \tag{23}$$

where

$$\underset{\sim}{\phi}(t') = \int_0^{t'} e^{(\underset{\sim}{A}-\nu I)(t'-t'')}\, \underset{\sim}{B}\, E\{\underset{\sim}{Y}(t'')\}dt''$$

since then $E\{\Delta(t')\Delta(t'')\underset{\sim}{Y}(t'')\} = E\{\Delta(t')\Delta(t'')\}E\{\underset{\sim}{Y}(t'')\} = e^{-\nu|t'-t''|}E\{\underset{\sim}{Y}(t'')\}$. Equation (23) was then solved using Laplace transform techniques and the stability in the case of a delta correlated Δ was investigated.

Note that alternatively, one could solve (20) using truncated hierarchy techniques [32] or solve (21) using successive approximations [5] or other techniques.

Smeach and Rust [34] used mean square theory and some results of Soong [36, Section 8.3.1] to study the linearized and randomized version of the system (cf. [35])

$$\frac{dx_1}{dt} = (r_1 - 1)\, x_1(t) - \frac{r_1 - 1}{x_{1e}}\, x_1^2(t) - c\, x_1(t)\, x_2(t)$$

$$\frac{dx_2}{dt} = \frac{r_2}{x_{1e}}\, x_1(t)\, x_2(t) - x_2(t)$$

obtained by retaining linear terms in the Taylor expansion about $x_{1S} = x_{1e}/r_2$, $x_{2S} = (r_1 - 1)(r_2 - 1)/c\, r_2$.

The system studied can be written as

$$\frac{d\underset{\sim}{X}}{dt} = \underset{\sim}{A}\,(\underset{\sim}{X}(t) - \underset{\sim}{X}_S)$$

where

$$\underset{\sim}{X} = \begin{bmatrix} X_1 \\ X_2 \end{bmatrix}, \quad \underset{\sim}{X}_S = \begin{bmatrix} X_{1S} \\ X_{2S} \end{bmatrix}, \quad \underset{\sim}{A} = \begin{bmatrix} \dfrac{R_1 - 1}{R_2} & -c\,\dfrac{x_{1e}}{R_2} \\ \dfrac{(R_1 - 1)(R_2 - 1)}{cx_{1e}} & 0 \end{bmatrix}$$

Its mean square solution is given by

$$\underset{\sim}{X}(t) = e^{\underset{\sim}{A}t}\,\underset{\sim}{X}(0) + \left\{\int_0^t e^{\underset{\sim}{A}(t'-t)}dt\right\}\underset{\sim}{A}\,\underset{\sim}{X}_S$$

and the joint probability density function of $\underset{\sim}{X}$ and $\underset{\sim}{R} = (R_1, R_2)^T$ is obtained by using Liouville's theorem [36, Section 8.3.1] as

$$f(\underset{\sim}{x}, \underset{\sim}{r}; t) = e^{-t\,\mathrm{Tr}\,\underset{\sim}{A}}\,\phi(\underset{\sim}{x}_S + e^{-\underset{\sim}{A}t}(\underset{\sim}{x}(t) - \underset{\sim}{x}_S), \underset{\sim}{r})$$

where $(\underset{\sim}{x}_0, \underset{\sim}{r})$ is the joint probability density function of $\underset{\sim}{X}_0 = \underset{\sim}{X}(0)$ and R.

Assuming $\underset{\sim}{X}_0$ and $\underset{\sim}{R}$ are independent the first moments of the solution are obtained as

$$E\{\underset{\sim}{X}(t)\} = E\{\underset{\sim}{X}_S\} - E\{e^{\underset{\sim}{A}t}\underset{\sim}{X}_S\} + E\{e^{\underset{\sim}{A}t}\} E\{X_0\}$$

$$\mathrm{Var}(\underset{\sim}{X}(t)) = \mathrm{Var}[\underset{\sim}{X}_S + (E\{\underset{\sim}{X}_0\} - \underset{\sim}{X}_S)e^{\underset{\sim}{A}t}] + E\{e^{\underset{\sim}{A}t}\underset{\sim}{V}e^{\underset{\sim}{A}t}t\}$$

where $\underset{\sim}{V}$ is the covariance matrix for $\underset{\sim}{X}_0$.

Numerical computations were carried out for various distributions of $\underset{\sim}{X}_0$ and $\underset{\sim}{R}$ and the results were compared to the solution of the corresponding deterministic system, (cf. [34] for details).

Bécus [6] studied system (19) making use of Soong's results [36, Section 8.3.1]. Rewriting (19) in vector form as

$$\frac{d\underset{\sim}{Y}}{dt} = \underset{\sim}{A}\,\underset{\sim}{Y} + \underset{\sim}{V}$$

where

$$\underset{\sim}{Y} = \begin{bmatrix} Y_1 \\ Y_2 \end{bmatrix}, \quad \underset{\sim}{A} = \begin{bmatrix} -\gamma_1 & -\beta_1 \\ \beta_2 & -\gamma_2 \end{bmatrix}, \quad \underset{\sim}{V} = \begin{bmatrix} \alpha_1 \\ -\alpha_2 \end{bmatrix}$$

and whose solution (subject to the initial condition $\underset{\sim}{Y}(0) = \underset{\sim}{Y}_0$) is

$$\underset{\sim}{Y}(t) = T_A(t)\,\underset{\sim}{Y}_0 + \int_0^t T_A(t - s)\underset{\sim}{V}\,ds \tag{24}$$

where $\{T_A(t),\ t \geq 0\}$ is the semigroup of random operators generated by $\underset{\sim}{A}$ and with the representation

$$T_A(t) = \exp\left[\int_0^t \underset{\sim}{A}(s)\,ds\right]$$

Taking expectation in (24) yields (assuming $\underset{\sim}{Y}_0$ is deterministic)

$$E\{\underset{\sim}{Y}(t)\} = E\{T_A(t)\}\underset{\sim}{Y}_0 + \int_0^t E\{T_A(t - s)\underset{\sim}{V}\}ds \tag{25}$$

For example if $\underset{\sim}{A}$ and $\underset{\sim}{V}$ are independent, $\underset{\sim}{V}$ being a random constant vector, and $\underset{\sim}{A} = \underset{\sim}{C}\, Z$ where $\underset{\sim}{C}$ is a constant deterministic matrix and Z is a random variable uniformly distributed in the interval $[Z_1, Z_2]$, $0 \leq Z_1 \leq Z_2$, one obtains from (25)

$$E\{\underset{\sim}{Y}(t)\} = \underset{\sim}{C}^{-1}[\exp(\underset{\sim}{C}\, Z_2 t)] - \exp(\underset{\sim}{C}\, Z_1\, t)]\underset{\sim}{Y}_0/t(Z_2 - Z_1)$$

$$+ \frac{\underset{\sim}{C}^{-1}}{(Z_2 - Z_1)} \int_0^t [\exp(\underset{\sim}{C}\, Z_2\, s) - \exp(\underset{\sim}{C}\, Z_1\, s)]s^{-1} ds\; E\{\underset{\sim}{V}\}$$

with $\underset{\sim}{C} = 2E\{\underset{\sim}{A}\}/(Z_1 + Z_2)$, which upon comparison with the solution

$$\overline{\underset{\sim}{Y}}(t) = \exp(E\{\underset{\sim}{A}\}t)\underset{\sim}{Y}_0 + \int_0^t \exp(E\{\underset{\sim}{A}\}s)ds\; E\{\underset{\sim}{V}\}$$

to the associated deterministic system shows that the expected solution of the random system is more stable than that of the deterministic one.

Alternatively, if $\underset{\sim}{A}$ and $\underset{\sim}{V}$ are random constants or random processes with finite degrees of randomness* we can rewrite system (19) as

$$\frac{d\underset{\sim}{Y}}{dt} = \underset{\sim}{h}(\underset{\sim}{Y}(t), \underset{\sim}{Z}(t), t) \qquad (26)$$

where $\underset{\sim}{Z}(t) = \underset{\sim}{Z}(\underset{\sim}{B}, t)$, $\underset{\sim}{B}$ being a finite, say n-dimensional, random vector. Liouville's theorem allows us to find the joint probability density function of $\underset{\sim}{Y}$ and $\underset{\sim}{B}$ as

$$f(\underset{\sim}{y}, \underset{\sim}{b}; t) = f_0(T_A(-t)[\underset{\sim}{y} - \underset{\sim}{w}], \underset{\sim}{b})\exp \int_0^t \text{tr}\; \underset{\sim}{A}(\tau)d\tau \qquad (27)$$

where $\underset{\sim}{W} = \int_0^t T_A(t - s)\underset{\sim}{V}\, ds$ and $f_0(\underset{\sim}{y}_0, \underset{\sim}{b})$ is the joint probability density

*A(t) is said to have finite degree of randomness if it is of the form $A(t) = g(t, \underset{\sim}{B})$ where $\underset{\sim}{B}$ is a constant finite random vector and g is a known deterministic function.

of $\underset{\sim}{Y}_0$ and $\underset{\sim}{B}$. From (27) the joint density of $\underset{\sim}{Y}$ can be found by integration as

$$f(\underset{\sim}{y}; t) = \int_{R^n} f(\underset{\sim}{y}, \underset{\sim}{b}; t) d \underset{\sim}{b} \tag{28}$$

For example when $\underset{\sim}{A}$ is deterministic and $\underset{\sim}{V}$ is a random constant vector then (27) yields

$$f(\underset{\sim}{y}, \underset{\sim}{b}; t) = \delta[\underset{\sim}{v} - (\exp(\underset{\sim}{A}t) - \underset{\sim}{I}) \underset{\sim}{A}(\underset{\sim}{y} - \exp(\underset{\sim}{A}t)\underset{\sim}{Y}_0)] \cdot$$

$$\cdot f(\underset{\sim}{y}) \exp[(\gamma_1 + \gamma_2)t]$$

In the absence of self interactions ($\gamma_1 \equiv \gamma_2 \equiv 0$) and with $\underset{\sim}{V}$ deterministic constant and $\underset{\sim}{A}$ random constant Eq. (27) yields

$$f(\underset{\sim}{y}, \underset{\sim}{b}; t) = \delta[\exp(-\underset{\sim}{A}t)(\underset{\sim}{y} - \underset{\sim}{A}^{-1}\exp(\underset{\sim}{A}t)\underset{\sim}{V} + \underset{\sim}{A}^{-1}\underset{\sim}{V}) - \underset{\sim}{Y}_0]f(\underset{\sim}{b}) \tag{29}$$

with $\underset{\sim}{Z} = \underset{\sim}{B} = [\beta_1, \beta_2]^T$.

If F is a smooth enough function we can write

$$E\{F(\underset{\sim}{Y})\} = \int F(\underset{\sim}{y}) f(\underset{\sim}{y}; t)d\underset{\sim}{y}$$

which upon using (28) becomes

$$E\{F(\underset{\sim}{Y})\} = \int F(\underset{\sim}{y}) \int_{R^n} f(\underset{\sim}{y}, \underset{\sim}{b}; t) d \underset{\sim}{b} d\underset{\sim}{y}$$

For example with $f(\underset{\sim}{y}, \underset{\sim}{b}; t)$ given by (29) we obtain

$$E\{F(\underset{\sim}{Y})\} = \int F[\exp(\underset{\sim}{A}t)\underset{\sim}{Y}_0 + \underset{\sim}{A}^{-1}\exp(\underset{\sim}{A}t)\underset{\sim}{V} - \underset{\sim}{A}^{-1}\underset{\sim}{V}]f(\underset{\sim}{b})d \underset{\sim}{b}$$

from which the first few moments of the solution can be obtained by taking $F(\underset{\sim}{Y}) = Y_i$, $i = 1, 2$ and $F(\underset{\sim}{Y}) = Y_1Y_2$.

Note that all the works using the random equation approach are based on models which can be linearized (Gompertz model) or on a linear approximation of nonlinear models as in [34].

RANDOM EVOLUTION MODELING

Bécus [7] used the formalism and results of the theory of random evolutions [16] to study the Gompertzian model (26). The stochastic process Z(t) is taken to be a Markov chain with state space {1,..., n}, stationary transition probabilities $p_{ij}(t)$ and infinitesimal matrix $\underset{\sim}{Q} = [q_{ij}] = p'_{ij}(0)]$. Thus the prey-predator system can evolve in any one of n modes specified by the evolution equations

$$\frac{d\underset{\sim}{Y}}{dt} = \underset{\sim}{h}_i\,(\underset{\sim}{Y}(t),\, t), \qquad t \in \{t > 0:\; Z(t) = i\}, \qquad i = 1,\dots, n$$

with $\underset{\sim}{h}_i = \underset{\sim}{h}|_{Z=i}$, $\underset{\sim}{h}$ being as defined in (26). The switching between these modes occurs at jump times of Z(t).

Let $\underset{\sim}{Y}_i$ denote the conditional expectation of $\underset{\sim}{Y}$ given that evolution started in state i, (i.e., Z(0) = i), i = 1,..., n then as an application of Theorem 2 in [16] we obtain

$$\frac{d\underset{\sim}{Y}_i}{dt} = \underset{\sim}{h}_i\,(\underset{\sim}{Y}_i(t),\, t) + \sum_j q_{ij}\,\underset{\sim}{Y}_j \tag{30}$$

In the case n = 2 (two modes of evolution) and with $\underset{\sim}{h}_i(\underset{\sim}{Y}, t) = \underset{\sim}{A}_i\underset{\sim}{Y} + \underset{\sim}{V}_i$ we obtain from (30) with Z a random telegraph signal process with parameter a (cf. [36], Ex. 3.6, for example)

$$\frac{d\underset{\sim}{U}}{dt} = \underset{\sim}{E}\,\underset{\sim}{U} + \underset{\sim}{W} \tag{31}$$

where

$$\underset{\sim}{U} = \left[\begin{array}{c} \underset{\sim}{Y}_1 \\ \hline \underset{\sim}{Y}_2 \end{array}\right], \qquad \underset{\sim}{E} = \left[\begin{array}{c|c} \underset{\sim}{A}_1 - \underset{\sim}{I}a & \underset{\sim}{I}a \\ \hline \underset{\sim}{I}a & \underset{\sim}{A}_2 - \underset{\sim}{I}a \end{array}\right], \qquad \underset{\sim}{W} = \left[\begin{array}{c} \underset{\sim}{V}_1 \\ \hline \underset{\sim}{V}_2 \end{array}\right]$$

Thus, in this case, switching between the two modes occur at jump times of a standard Poisson process with intensity a > 0 and the average time spent in each mode before switching is a^{-1}.

The behavior of the solution to (31) depends upon the nature of the eigenvalues of the matrix $\underset{\sim}{E}$, which leads to the study of the nature of the roots of a 4-th degree polynomial. The algebra is quite ponderous. With

$$\underset{\sim}{A}_i = \begin{bmatrix} -\gamma_{i1} & -\beta_{i1} \\ \beta_{i2} & -\gamma_{i2} \end{bmatrix}, \qquad i = 1, 2$$

and in the absence of self interactions ($\gamma_{ij} = 0$, i, j = 1, 2) the following result is proved:

(i) if $(\beta_{11} - \beta_{21})(\beta_{12} - \beta_{22}) > 0$ the solution is asymptotically stable;

(ii) if $(\beta_{11} - \beta_{21})(\beta_{12} - \beta_{22}) < 0$ the solution is

(a) asymptotically stable for

$$a^2 > a_c^2 = - \frac{(\beta_{11}\beta_{12} - \beta_{21}\beta_{22})^2}{4(\beta_{11} - \beta_{21})(\beta_{12} - \beta_{22})}$$

(b) oscillatory or asymptotically stable if $a^2 = a_c^2$,

(c) possibly unstable if $a^2 < a_c^2$;

(iii) if $(\beta_{11} - \beta_{21})(\beta_{12} - \beta_{22}) = 0$ the solution is asymptotically stable unless $\beta_{11} = \beta_{21}$ and $\beta_{12} - \beta_{22}$ in which case it is oscillatory.

These results can be contrasted with those for the associated deterministic system

$$\frac{d\underset{\sim}{\overline{Y}}}{dt} = \frac{\underset{\sim}{A}_1 + \underset{\sim}{A}_2}{2} \underset{\sim}{\overline{Y}} + \frac{\underset{\sim}{V}_1 + \underset{\sim}{V}_2}{2} \tag{32}$$

whose solution is always oscillatory.

As the random evolution model leads to algebraic complexity, the feasibility of using available diffusion approximations was also investigated in [7] with the following results.

Introducing a small parameter $\varepsilon = (\varepsilon_1/\varepsilon_2)^2$ in (30) we obtain

$$\frac{d\underset{\sim}{Y}_i^{\varepsilon}}{d\tau} = \frac{1}{\sqrt{\varepsilon}} \underset{\sim}{h}_i(\underset{\sim}{Y}_i^{\varepsilon}, \tau) + \frac{1}{\varepsilon} \sum q_{ij} Y_j^{\varepsilon}, \qquad i = 1, 2$$

where $\tau = (\varepsilon_1^2/\varepsilon_2)t$. It can be shown that as $\varepsilon \to 0$, $\tilde{Y}_i^\varepsilon$ converges to the solution of

$$\frac{d\tilde{Y}^o}{d\tau} = \frac{1}{2}\left\{\sum_{i,j} c_{ij} (\tilde{h}_i - \sum p_k \tilde{h}_k)(\tilde{h}_j - \sum p_k \tilde{h}_k)\right\}\tilde{Y}^o \tag{33}$$

where

$$c_{ij} = \lim_{\tau\to\infty} \frac{1}{\tau} \operatorname{Cov}(\gamma_i(\tau)\ \gamma_j(\tau))$$

$\gamma_i(\tau)$ being the occupation time in state i up to time τ, and p_i is the unique solution of

$$\sum p_i q_{ij} = 0, \quad \sum p_i = 1$$

In the case when Z is a random telegraph signal process with parameter a and $\tilde{h}_i(\tilde{Y}) = \tilde{A}_i \tilde{Y} + \tilde{V}_i$, Eq. (33) reduces to

$$\frac{d\tilde{Y}^o}{d\tau} = (8a)^{-1} [(\tilde{A}_1 - \tilde{A}_2)^2\tilde{Y}^o + (\tilde{A}_1 - \tilde{A}_2)(\tilde{V}_1 - \tilde{V}_2)]$$

The behavior of its solution depends upon the nature of the eigenvalues of $(\tilde{A}_1 - \tilde{A}_2)^2$. In the absence of self interaction we recover our earlier result: if $(\beta_{11} - \beta_{21})(\beta_{12} - \beta_{22}) > 0$ we have asymptotic stability. In the general case we have asymptotic stability if $(\gamma_{21} - \gamma_{11})^2 + (\gamma_{22} - \gamma_{12})^2 - 2(\beta_{21} - \beta_{11})(\beta_{22} - \beta_{12}) < 0$.

Another diffusion approximation is obtained from

$$\frac{d\tilde{Y}_i^\varepsilon}{dt} = \tilde{h}_i(\tilde{Y}_i^\varepsilon, t) + \frac{1}{\varepsilon}\sum q_{ij} Y_j^\varepsilon, \qquad i = 1, 2$$

and shows that as $\varepsilon \to 0$, Y_i^ε converges to $\tilde{\overline{Y}}$ the solution to (32).

Note that the random evolution model is well suited to describe a situation in which the prey predator system, in the absence of random effects, would evolve in a periodic fashion due for example to periodic (e.g., seasonal) changes in the environment. The random evolution model accounts for random breakdowns in this periodicity. Thus another associated deterministic system would be

$$\frac{d\underset{\sim}{Y}}{dt} = \underset{\sim}{h}(\underset{\sim}{Y}(t), t)$$

with $\underset{\sim}{h}$ periodic and which could be studied in the context of Floquet theory.

MISCELLANEOUS RESULTS

We assemble here a brief review of various results on stochastic prey-predator systems which do not quite fit any of the previous headings.

Billard [10] considered the evolution of a prey predator system in $A = \{0 \le X_i \le X_i(0), i = 1, 2\}$ for which losses in the prey population occur through predation and losses in the predator population result from natural attrition. She obtained the Chapman-Kolmogorov equation for the joint transition probability distribution

$$\begin{aligned}\frac{dP(x_1, x_2; t)}{dt} &= -x_2(\mu_1 x_1 + \mu_2) P(x_1, x_2; t) \\ &+ \mu_1(x_1 + 1)x_2 P(x_1 + 1, x_2; t) \\ &+ \mu_2(x_2 + 1) P(x_1, x_2 + 1; t), \qquad (x_1, x_2) \in A \qquad (34)\end{aligned}$$

with the obvious condition $P(x_1, x_2; t) = 0$ if $(x_1, x_2) \notin A$.

Rather than transform (34) into a partial differential equation by introducing a generating function (cf. second and fourth sections) she used a technique of Severo [33] to solve the differential-difference equation (34). The solution is of the form

$$P(x_1, x_2; t) = \sum_{m,w} C_{N_2-x_2+1,\, m}(N_1 - x_1 + 1, w) \exp[b_{mm}(w, w)]$$

where $N_i = X_i(0)$, $i = 1, 2$ and the coefficients $b_{mm}(w, w)$ and $C_{\ell m}(k, w)$ are given in terms of the initial population sizes N_i and rate constants μ_i.

Goh and Jennings [15] considered a random assemblage of Lotka-Volterra models each of which has a feasible equilibrium. They found that it has the same stability property as a set of linear models (obtained from the Lotka-Volterra model

$$\frac{dx_i}{dt} = x_i[b_i + a_{ij} x_j], \qquad i = 1, 2, \ldots, m$$

by letting $x_i = \bar{x}_i + y_i$, where $(\bar{x}_1, \bar{x}_2, \ldots, \bar{x}_m)$ is the nontrivial equilibrium, and keeping linear terms in y_i yielding

$$\frac{dy_i}{dt} = \sum x_i a_{ij} y_j, \qquad i = 1, 2, \ldots, m)$$

assembled in the same random fashion. Also of the set of random assemblages of Lotka-Volterra models only a fraction behaving as 2^{-m} consists of models with feasible equilibrium. These results support May's contention that stability decreases with diversity.

Turelli [39] discussed the various concepts of stability for stochastic models in population biology. In particular he showed that May's criterion (15) is valid if randomness enters only through the intrinsic growth rates of the populations. On the other hand, Yodzis [41] introduced the notion of tenacity which is the mean time for the first passage of the system out of the domain of attraction of the equilibrium, starting from the equilibrium and related it to May's criterion (15) thus providing a more precise meaning to May's concept of stochastic stability.

CONCLUSIONS

In spite of the large body of literature available, much remains to be done in stochastic modeling and analysis of prey-predator systems.

The nonlinearity in the interactions become especially difficult to handle in a stochastic model. For example all the works surveyed in the sixth section are restricted to linear or linearized models. It is hoped that as more results on nonlinear random differential equations become available researchers will tackle nonlinear models.

The results of the stochastic analyses surveyed are contradictory. Some works indicate that randomness has a stabilizing effect while others claim a destabilizing effect. Perhaps much confusion stems from the use of the word "stable" with different meanings; this certainly appears to be the case in the controversy surrounding May's criterion.

I am not aware of any stochastic model including spatial dependency and/or aging effects, and/or time delays. These certainly are areas in which much research is needed.

REFERENCES

1. V. D. Barnett, "The Monte Carlo solution of a competing species problem," *Biometrics* 18(1962), 76-103.

2. M. S. Bartlett, "On theoretical models for competitive and predatory biological systems," *Biometrika* 44(1957), 27-42.

3. M. S. Bartlett, *Stochastic Population Models in Ecology and Epidemiology*. Methuen, London (1960).

4. N. G. Becker, "A stochastic model for two interacting populations," *J. Appl. Prob.* 7(1970), 544-564.

5. G. A. Bécus, "Successive approximation solutions of a class of random equations," in *Approximate Solution of Random Equations*, A. T. Bharucha-Reid (Ed.). North Holland (1979), 1-12.

6. G. A. Bécus, "Stochastic prey-predator relationships: a random differential equation approach," *Bull. Math. Biol.* 41(1979), 91-100.

7. G. A. Bécus, "Stochastic prey-predator relationships: a random evolution approach," *Bull. Math. Biol.* 4(1979), 543-554.

8. A. T. Bharucha-Reid, *Random Integral Equations*. Academic Press, New York (1972).

9. A. T. Bharucha-Reid, *Elements of the Theory of Markov Processes and Their Applications*. McGraw-Hill, New York (1960).

10. L. Billard, "On Lotka-Volterra predator prey models," *J. Appl. Prob.* 14(1977), 375-381.

11. C. L. Chiang, "Competition and other interactions between species," in *Statistics and Mathematics in Biology*. Iowa State University Press, Ames, Iowa (1954), 197-215.

12. T. C. Gard and D. Kannan, "On a stochastic differential equation modeling of prey-predator evolution," *J. Appl. Prob.* 13(1976), 429-443.

13. N. S. Goel and N. Richter-Dyn, *Stochastic Models in Biology*. Academic Press, New York (1974).

14. N. S. Goel, S. C. Maitra and E. W. Montroll, "On the Volterra and other nonlinear models of interacting populations," *Rev. Modern Phys.* 43 (1971), 231-276.

15. B. S. Goh and L. S. Jennings, "Feasibility and stability in randomly assembled Lotka-Volterra models," *Ecological Modelling* 3(1977), 63-71.

16. R. Griego and R. Hersh, "Theory of random evolutions with applications to partial differential equations," *Trans. Amer. Math. Soc.* 156(1971), 405-418.

17. M. Iosifescu and P. Tautu, *Stochastic Processes and Applications in Biology and Medicine II*. Springer-Verlag, Berlin (1973).

18. D. Kannan, "On some Markov models of certain interacting populations," *Bull. Math. Biol.* 38(1976), 723-738

19. A. Kolmogorov, "Sulla teoria di Volterra della lotta per l'esistenza," *G. Istit. Ital. Degli Attuari* 7(1936), 74-80.

20. Kostitzin, *Symbiose, Parasitisme et Evolution (Etude Mathêmatique)*. Hermann & Cie, Paris (1934).

21. P. H. Leslie and J. C. Gower, "The properties of a stochastic model for two competing species," *Biometrika* 45(1958), 316-330.

22. P. H. Leslie and J. C. Gower, "The properties of a stochastic model for the predator-prey type of interaction between two species," *Biometrika* 47(1960), 219-234.

23. P. H. Leslie, "A stochastic model for studying the properties of certain biological systems by numerical methods," *Biometrika* 45(1958), 16-31.

24. A. J. Lotka, *Elements of Physical Biology*. Williams & Wilkins, Baltimore, Maryland (1925).

25. R. M. May, *Stability and Complexity in Model Ecosystems*. Princeton University Press, Princeton, New Jersey (1973).

26. R. M. May, "Stability in randomly fluctuating versus deterministic environments," *Amer. Natur.* 107(1973), 621-650.

27. J. B. Morton and S. Corrsin, "Experimental confirmation of the applicability of the Fokker-Planck equation to a non-linear oscillator," *J. Math. Phys.* 10(1969), 361-368.

28. R. W. Poole, "A discrete time stochastic model of a two prey one predator species interaction," *Theor. Pop. Biol.* 5(1974), 208-228.

29. Prajneshu, "A stochastic model for two interacting species," *Stoch. Proc. Applic.* 4(1976), 271-282.

30. P. S. Puri, "A linear birth and death process under the influence of another process," *J. Appl. Prob.* 12(1975), 1-17.

31. A. Rescigno and I. W. Richardson, "The deterministic theory of population dynamics," in *Foundations of Mathematical Biology*, vol. III, R. Rosen (Ed.), Academic Press, New York (1973).

32. J. M. Richardson, "The application of truncated hierarchy techniques in the solution of a stochastic linear differential equation," *Proc. Sympos. Appl. Math, XVI*, Amer. Math. Soc. Providence, Rhode Island, (1964), 290-302.

33. N. C. Severo, "A recursion theorem on solving differential-difference equations and applications to some stochastic processes," *J. Appl. Prob.* 6(1969), 673-681.

34. S. C. Smeach and A. Rust, "A stochastic approach to predator-prey models," *Bull. Math. Biol.* 40(1978), 483-498.

35. J. M. Smith, *Mathematical Ideas in Biology*. Cambridge University Press, London (1968).

36. T. T. Soong, *Random Differential Equations in Science and Engineering*. Academic Press, New York (1973).

37. W. C. Torrez, "The birth and death chain in a random environment: instability and extinction theorems," *Ann. of Prob.* 6(1978), 1020-1043.

38. M. Turelli, "Random environments and stochastic calculus," *Theor. Pop. Biol.* 12(1977), 140-178.

39. M. Turelli, "A reexamination of stability in randomly varying versus deterministic environments with comments on the stochastic theory of limiting similarity," *Theor. Pop. Biol.* 13(1978), 244-267.

40. V. Volterra, "Variazoni e fluttuazioni del numero d'individui in specie animali conviventi," *Mem. R. Comitato Talassografico Ital.* 131(1927), 1-142.

41. P. Yodzis, "Environmental randomness and the tenacity of equilibria," *J. Theor. Biol.* 72(1978), 185-189.

COEXISTENCE IN PREDATOR-PREY SYSTEMS

G. J. Butler

Department of Mathematics
University of Alberta
Edmonton, Alberta, Canada

Of some interest in the study of predator-prey systems is the possibility of the coexistence of some number of distinct predatory species in competition for a food source consisting of some smaller number of species of prey. In studies related to the observation of plankton populations, MacArthur [12] and Stewart and Levin [16] suggested that under appropriate circumstances coexistence could occur. MacArthur expressed this possibility in terms of one species more suited to growing at low resource levels coexisting competitively with another species better adapted for growing at higher resource densities.

Formally we identify a predator-prey system as a system of k prey species with time dependent population densities $S_1(t),\ldots, S_k(t)$, and n predator species with population densities $x_1(x),\ldots, x_n(t)$, dynamically related by a system of ordinary differential equations

$$\begin{cases} S_i' = g_i(S_i) - \sum_{j=1}^{n} p_{ij}(S_i, x_j), & i = 1,\ldots, k \\ \\ x_j' = \sum_{i=1}^{k} q_{ij}(S_i, x_j) - d_j(x_j), & j = 1,\ldots, n \end{cases} \qquad (P)$$

Here g_i, d_j are, respectively, the growth rate of the i-th prey in the absence of predation and the death rate of the j-th predator in the absence of prey. p_{ij} is the predation response of the j-th predator to the i-th prey and q_{ij} is the conversion of this response into predator growth. For mathematical convenience, all functions are considered to be C^1 functions of their (nonnegative) arguments. In order to impart the basic predator-prey characteristics to the model, it is assumed that each function vanishes when its argument (or either of its arguments) vanishes, that d_j is positive and monotone increasing for $x_j > 0$, that p_{ij} and q_{ij} are positive and monotone increasing in each argument for S_i, $x_j > 0$ and that g_i is positive and increasing for small positive values of S_i (it may become negative if there is assumed a carrying capacity for the i-th prey).

Since our interest here is in competition for a limited number of resource types, we assume $k < n$ throughout this paper. While this model is certainly not as general as it could be, for example ignoring stochastic effects, time delays and allowing no interaction between distinct prey species nor (other than direct competition for resources) between distinct predator species, it does allow the inclusion of the more salient aspects of predator-prey interaction.

Under the present assumptions, the equations may be written

$$S_i' = S_i G_i(S_i, x_1,\ldots, x_n), \qquad i = 1,\ldots, k$$

$$x_j' = x_j H_j(S_1,\ldots, S_k, x_j), \qquad j = 1,\ldots, n$$

where G_i, H_j are continuous functions termed the specific growth rates of the i-th prey, j-th predator, respectively. The positive cone $C_{k,n}$ of $\mathbb{R}^k \times \mathbb{R}^n$, where $C_{k,n} = \{(S_1,\ldots, S_k, x_1,\ldots, x_n): S_i > 0, x_j > 0, i = 1,\ldots, k; j = 1,\ldots, n\}$, is invariant for the system (P) as is each of its coordinate plane boundaries, obviously a necessary requirement for the realism of the model. We shall denote the restriction of (P) to $C_{k,n}$ by (P^+).

With some additional reasonably realistic assumptions, we could guarantee that all solutions of (P) are uniformly bounded as $t \to \infty$. For example, it is enough to assume

(a) there is a carrying capacity for each species, i.e., $g_i(S_i) < 0$ for all $S_i > K_i$, for some $K_i > 0$, $i = 1,\ldots, k$.

(b) $d_j(x_j) \to \infty$ as $x_j \to \infty$ for each j

(c) there exist μ_{ij}, $i = 1,\ldots, k$; $j = 1,\ldots, n$, such that $q_{ij}(S_i, x_j) \le \mu_{ij}p_{ij}(S_i, x_j)$ for all $S_i, x_j > 0$. This last condition is natural as it bounds the rate of resource conversion (to predator growth) to the rate of resource depletion.

Following McGehee and Armstrong [14], we shall say that (P) persists (exhibits coexistence) if $C_{k,n}$ contains a compact attractor for (P).

The absence of coexistence for a system is tied up with the idea of a principle of competitive exclusion, and the early examples considered seemed to enhance such a principle. These examples derived from assuming equations of Lotka-Volterra type, i.e.,

$$\begin{cases} S_i' = S_i\left[c_i - K_iS_i - \sum_{j=1}^{n} \ell_j x_j\right], & i = 1,\ldots, k \\ x_j' = x_j\left[\sum_{i=1}^{n} m_iS_i - d_j\right], & j = 1,\ldots, n \end{cases} \qquad (LV)$$

where c_i, m_i, ℓ_j, d_j are positive constants and K_i is a nonnegative constant. Note that the specific growth rate for each species is an affine function. Levin [11] showed that (LV) can never possess an asymptotically stable critical point nor a stable periodic orbit within $C_{k,n}$, and then McGehee and Armstrong [14] showed that in fact any system (P) for which the specific predator growth rates are affine must fail to persist. In the same paper they demonstrated, however, that coexistence could occur if nonlinear specific growth rates were allowed, by constructing such a system with $k = 1$, $n = 2$. Since then Zicarelli [17] has constructed persistent systems for $k = 1$, n arbitrary, and Grasman [4] has a $k = 1$, $n = 2$ system

system which possesses a periodic attractor in $C_{1,2}$. We should remark that coexistence for systems (P) is just part of a more general question concerning coexistence for consumer-resource systems. See [1, 2, 10, 15].

Since the question of whether there can be systems (P) exhibiting coexistence is answered, we turn our attention to systems which incorporate one of the specific forms of predator response that have been suggested in the literature (see, for example [13]). One such response is that considered by Holling in [5] and studied numerically by Koch [9] and analytically by Hsu, Hubbell and Waltman [6, 7] for $k = 1$, $n = 2$. Since we wish to discuss this particular model in detail, we reproduce the argument behind its formulation.

For a given predator and a given prey species, let T be the total attack time, T_s the search time, h the "handling" time per unit prey caught and N_a the number of prey caught during the attack cycle. We have

$$T = T_s + hN_a$$

Assuming the number of prey caught to be proportional to prey density S and search time, we also have

$$N_a = cT_sS$$

where c is a constant (the encounter rate per unit prey density). The above equations yield the feeding rate per predator

$$F = \frac{N_a}{T} = \frac{cS}{1 + chS}$$

and the total consumption of the prey population by the predator population occurs at a rate FX where X is the predator density. Assuming logistic growth for the prey, the above Holling form of response for each of the predators, and constant specific death rates for the predators, the model becomes

$$S' = \gamma S\left[1 - S/K - \frac{m_1}{y_1}\frac{x_1}{a_1 + S} - \frac{m_2}{y_2}\frac{x_2}{a_2 + S}\right]$$

$$x_1' = x_1\left[\frac{m_1 S}{a_1 + S} - D_1\right] \qquad (H_{1,2})$$

$$x_2' = x_2\left[\frac{m_2 S}{a_2 + S} - D_2\right]$$

All of the parameters in this system are positive constants. γ is the maximum specific growth rate for the prey, K is its carrying capacity, D_i is the specific death rate and m_i is the maximum specific birth rate for the i-th predator. a_i is the "half-saturation" constant for the i-th predator. Relating these parameters back to the derivation of the Holling type response, we find that m_i is proportional to the inverse of the handling time h and a_i is proportional to the inverse of the handling time and the encounter rate c.

Koch's numerical work suggested coexistence for quite a large range of values of the parameter space.

In order to summarize the results of Hsu, Hubbell and Waltman, it is necessary to describe briefly the characteristics of the one-predator-one prey system which is the restriction of $(H_{1,2})$ to the (S, x_i)-plane (i = 1, 2). Denote the restrictions by $(H_{1,1(i)})$ and the corresponding positive cone by $C_{1,1(i)}$. Then

(i) If the maximum birth rate is too small, i.e., if $b_i = m_i/D_i \leq 1$, or if $b_i > 1$ but the carrying capacity is too small, i.e., if $K \leq \lambda_i$, where $\lambda_i = a_i/(b_i - 1)$, then $(H_{1,1(i)})$ has two critical points, namely (0, 0, 0) which is unstable and (K, 0, 0) which is globally asymptotically stable for $(H^+_{1,1(i)})$.

(ii) if $b_i > 1$ and $\lambda_i < K$, then there is a critical point $E_i = (\lambda_i, x_i^*, 0)$ within $C_{1,1(i)}$, where $x_i^* = y_i\gamma(1 - \lambda_i/K)(a_i + \lambda_i)/m_i$. If $\lambda_i < K < a_i + 2\lambda_i$, then E_i is globally asymptotically stable for $(H^+_{1,1(i)})$, and we denote E_i by A_i. If $K > a_i + 2\lambda_i$, then E_i is unstable and $(H^+_{1,1(i)})$ has an attractor A_i within $C_{1,1(i)}$ whose boundary (boundaries) is a stable limit cycle (two semi-stable periodic orbits). A_i is global for $(H^+_{1,1(i)})$ with the exception of the single orbit E_i.

Returning to the system $(H_{1,2})$, the main results are as follows:

1. If for i = 1 or 2, we have $b_i = m_i/D_i \leq 1$ or $\lambda_i > K$, then for each solution of $(H^+_{1,2})$, we have $\lim_{t\to\infty} x_i(t) = 0$ and the attractor A_{3-i} for the restriction $(H^+_{1,1(3-i)})$ is also an attractor for $(H^+_{1,2})$; global if E_{3-i} is stable; global except for an exceptional orbit if E_{3-i} is unstable.

2. If $b_i > 1$ and $\lambda_i < K$, i = 1, 2, so that each predator could survive in the absence of competition, then assume that $\lambda_1 < \lambda_2$. If $b_1 \geq b_2$, we have $\lim_{t\to\infty} x_2(t) = 0$ and have the same conclusion as in case 1 with regard to the attractor A_1.

3. If $1 < b_1 < b_2$, $\lambda_1 < \lambda_2 < K$, and $K < (a_2b_1 - a_1b_2)/(b_2 - b_1)$, the conclusion is as in case 2.

4. There remains the parameter region $1 < b_1 < b_2$, $\lambda_1 < \lambda_2 < K$, $K > (a_2b_1 - a_1b_2)/(b_2 - b_1)$.

In this case, the dynamical behavior of $(H_{1,2})$ was not resolved in [6, 7], but numerical work of the authors suggested that if the critical point E_1 was unstable, coexistence would occur for a range of values of K, thus reinforcing the observation of Koch. This author has verified this conjecture and we present a brief outline of the argument involved:

When case 4 applies, it may be shown that if E_2 is stable for $(H_{1,1(2)})$, then for all solutions of $(H^+_{1,2})$, we have $\lim_{t\to\infty} x_1(t) > 0$, $\lim_{t\to\infty} S(t) > 0$. Suppose now that E_1 is unstable and for some solution of $(H^+_{1,2})$ we have $\lim_{t\to\infty} x_2(t) = 0$. If this solution is any one other than that lying in the 1-dimensional stable manifold of E_1 (relative to $(H_{1,2})$), then its positive omega limit set Ω will intersect the attractor A_1 for $(H^+_{1,1(1)})$. This means that Ω contains some periodic orbit of $(H^+_{1,1(1)})$. Let the set of functions S(t) corresponding to a periodic orbit of $(H^+_{1,1(1)})$ be denoted by Σ. It can be shown that if the parameters of the system $(H_{1,2})$ are suitably chosen then

$$\lim_{t\to\infty} \left(\frac{1}{t} \int_0^t \frac{m_2\, S(\tau)}{a_2 + S(\tau)}\, d\tau \right) - D_2$$

will be positive and uniformly bounded away from zero, for all $S(t) \in \Sigma$. The result of this is that the attractor A_1 for $(H^+_{1,1(1)})$ is a repellor in the positive x_2-direction as regards the system $(H_{1,2})$. But that contradicts $A \cap \Omega \neq \emptyset$.

Thus, except for an exceptional orbit in which $S \to \lambda_1$, $x_1 \to x^*_1$ and $x_2 \to 0$ as $t \to \infty$, all solutions of $(H^+_{1,2})$ satisfy $\lim_{t\to\infty} x_i(t) > 0$, $i = 1, 2$, $\lim_{t\to\infty} S(t) > 0$. One may then deduce the existence of an attractor in $C_{1,2}$ for $(H^+_{1,2})$, global except for the singular orbit.

The details of the above argument will be found in [3].

Much the same analysis will work for some models based on other predation functions. For example, if we modify the Holling functional response by allowing the contact rate c to be a density dependent c(S), this is the case. As another example, using Ivlev's functional response $x(1 - e^{-S/a})$ [8], we obtain the following one prey-two predator system:

$$S' = \gamma S(1 - S/K) - \frac{m_1}{y_1}(1 - e^{-S/a_1}) - \frac{m_2}{y_2}(1 - e^{-S/a_2})$$

$$x_1' = x_1[m_1(1 - e^{-S/a_1}) - D_1] \qquad (I_{1,2})$$

$$x_2' = x_2[m_2(1 - e^{-S/a_2}) - D_2].$$

Again there exists a range of parameters for which $(I_{1,2})$ exhibits coexistence.

It is interesting to note that in the above one prey-two predator models coexistence occurs in parameter regions which correspond to one of the predators being more suitably adapted to growth at low resource densities, the other more adapted to high resource densities (see [7] for a discussion of this for the system $(H_{1,2})$) and it seems worthwhile to explore this possibility further.

Apart from the problem of characterizing persistent competitive systems, there are a number of interesting, related questions, of which we give a sample:

1. When can the attractor of coexistence be shown to be a limit cycle? There seems to be good numerical evidence for this type of behavior for the system $(H_{1,2})$ [7].

2. In models which exhibit coexistence for a certain range of the parameters involved, when can the parameters be varied outside this range so that predominance passes from one of the competing species to another?

3. What is the effect on a persistent system of allowing "interference" between the competitors?

Finally, one would like to study the effect of time delays and time-dependent parameters on the property of coexistence.

ACKNOWLEDGMENT

This research was supported in part by the National Research Council of Canada grant NRC A-8130.

REFERENCES

1. R. A. Armstrong and R. McGehee, "Coexistence of two competitors on one resource," *J. Theor. Biol.* 56(1976), 499-502.

2. R. A. Armstrong and R. McGehee, "Coexistence of species competing for shared resources," (to appear).

3. G. J. Butler, "The transition from extinction to coexistence in a predator-prey model," (to appear).

4. W. Grasman, "The existence of a periodic solution to a model of two predators and one prey," Presented to the NSF-CBMS Conference on Modelling and Differential Equations in Biology, Carbondale, Illinois (1978).

5. C. S. Holling, "The functional response of predators to prey density and its role in mimicry and population regulation," *Mem. Entomol. Soc. Canada* 45(1965), 5-60.

6. S. B. Hsu, S. P. Hubbell and P. Waltman, "Competing predators," *SIAM J. Appl. Math.* 35(1978), 617-625.

7. S. B. Hsu, S. P. Hubbell and P. Waltman, "A contribution to the theory of competing predators," *J. Math. Biol.* 48 (1978), 337-349.

8. V. S. Ivlev, *Experimental Ecology of the Feeding of Fishes*. Yale University Press, New Haven, Connecticut (1961).

9. A. L. Koch, "Competitive coexistence of two predators utilizing the same prey under constant environmental conditions," *J. Theor. Biol.* 44(1974), 378-386.

10. J. Kaplan and J. Yorke, "Competitive exclusion and nonequilibrium coexistence," *Amer. Natur.* 111(1977), 1030-1036.

11. S. Levin, "Community equilibria and stability, and an extension of the competitive exclusion principle," *The American Naturalist* 104, no. 939 (1970), 413-423.

12. R. H. MacArthur, "Population ecology of some warblers of northeastern coniferous forests," *Ecology* 39(1958), 599-619.

13. R. May, *Stability and Complexity in Model Ecosystems*. Princeton University Press, Princeton, New Jersey (1973).

14. R. McGehee and R. A. Armstrong, "Some mathematical problems concerning the ecological principle of competitive exclusion," *J. Differential Equations* 23(1977), 33-52.

15. Z. Nitecki, "A periodic attractor determined by one function," (to appear).

16. F. M. Stewart and B. R. Levin, "Partitioning of resources and the outcome of interspecific competition: a model and some general considerations," *Amer. Nat.* 107(1973), 171-198.

17. J. Zicarelli, *Mathematical analysis of a population model with several predators on a single prey*. Ph.D. thesis, University of Minnesota, Minneapolis, Minnesota (1975).

STABILITY OF SOME MULTISPECIES POPULATION MODELS

B. S. Goh

Mathematics Department
University of Western Australia
Nedlands, Australia

INTRODUCTION

Stability in a nonlinear population model is often established by examining the eigenvalues of its linearized dynamics. This method gives only stability relative to infinitesimal perturbations of the initial state.

But populations in the real world are subjected to large perturbations. It is therefore essential that a population model should be stable relative to finite perturbations of its initial state.

The direct method of Liapunov (see [4]) is a powerful analytical method for establishing stability relative to finite perturbations. This method requires the construction of a continuous and positive definite function which will be denoted by V(N). The time derivative of V(N) along every solution of the system in a finite region is nonpositive. Because of the fact that population variables are nonnegative, it is desirable for V(N) to tend to infinity as a population variable tends to zero. By definition a positive equilibrium is said to be globally stable if every solution which begins in the positive orthant remains in it and converges to the equilibrium as the time variable tends to infinity.

Consider a multispecies system in which each species is self-regulating and the interspecific interactions are relatively weak. For this type of multispecies model, a positive linear combination of Liapunov functions for self-regulating single species models is a natural candidate to act as a Liapunov function.

If one or more species in a multispecies is not self-regulating, then, as demonstrated by Hsu [3], it may be worthwhile to subject a Lotka-Volterra model to Poincaré transformation before using the direct method of Liapunov. The Poincaré transformation is applied repeatedly until every component in the transformed model is self-regulating or is neutrally stable on its own. Here the Poincaré transformation is applied to a general class of nonlinear population models.

SINGLE SPECIES MODEL

Let N be the density of a population in a region. In a model of the population, N will be treated as a continuous variable. Let F(N) be a continuous function on $R_+ = \{N|N > 0\}$. A general model of the population is

$$\dot{N} = NF(N) \tag{1}$$

By definition N^* is an equilibrium of (1) if $N^*F(N^*) = 0$. We say that a positive equilibrium N^* which satisfies the equation $F(N^*) = 0$ is globally asymptotically stable if every solution of (1) which begins in the set R_+ remains in it for all $t \geq 0$ and converges to N^* as $t \to \infty$.

The direct method of Liapunov is a powerful analytical method for establishing the global asymptotic stability of an equilibrium. For a single species model this method (see [1]) requires the construction of a continuous function V(N) which has the following properties: (i) V(N) is positive definite such that $V(N^*) = 0$. (ii) $V(N) \to \infty$ as $N \to 0+$ and as $N \to \infty$. (iii) $\dot{V}(N) = (\partial V/\partial N)NF(N)$ is nonpositive for all positive values of N.

The positive equilibrium N^* of (1) is globally asymptotically stable if there exists a Liapunov function $V(N)$ such that $\dot{V}(N)$ does not vanish identically along a nontrivial solution $N(t) \neq N^*$.

In a private communication S. B. Hsu suggested that a general expression for a Liapunov function of (1) is

$$V(N) = \int_{N^*}^{N} \frac{h(s)}{g(s)}\, ds \tag{2}$$

where $h(s)$ and $g(s)$ are continuous functions such that $h(s) < 0$ for all $s \in (0, N^*)$, $h(N^*) = 0$, $h(s) > 0$ for all $s \in (N^*, \infty)$ and $g(s) > 0$ for all $s > 0$. Furthermore $g(s)$ and $h(s)$ are chosen so that $V(N) \to \infty$ as $N \to 0+$ and as $N \to \infty$. Hsu [2] has used this type of Liapunov function in an analysis of a prey-predator model.

EXAMPLE 1 Let $h(s) = s - N^*$ and $g(s) = s$. We get $V(N) = N - N^* - N^* \ell n(N/N^*)$ and $\dot{V}(N) = (N - N^*)F(N)$. It follows that the positive equilibrium N^* of (1) is globally asymptotically stable if $F(N) < 0$ for all $N \in (0, N^*)$ and $F(N) > 0$ for all $N \in (N^*, \infty)$.

MULTISPECIES MODELS

Let N_i be the population of the ith species in a community of m interacting species. Suppose the community can be described by the model,

$$\dot{N}_i = N_i F_i(N_1, N_2, \ldots, N_m), \qquad i = 1, 2, \ldots, m \tag{3}$$

where $F_1, F_2, \ldots, F_m$ are continuous functions in the positive orthant $R_+^m = \{N | N_i > 0,\ i = 1, 2, \ldots, m\}$.

Let N^* denote an equilibrium which satisfies the equations $F_i(N) = 0$ for $i = 1, 2, \ldots, m$. We say that N^* is a positive equilibrium if $N_i^* > 0$ for $i = 1, 2, \ldots, m$.

For the population model in (3), a continuous function $V(N)$ is a Liapunov function if: (i) $V(N)$ is positive definite such that $V(N^*) = 0$,

(ii) $V(N) \to \infty$ as $N_i \to 0+$ and as $N_i \to \infty$ for $i = 1, 2, \ldots, m$, and (iii) the time derivative of $V(N)$ along every solution of (3) is nonpositive. We have

$$\dot{V}(N) = \sum_{i=1}^{m} \left(\frac{\partial V}{\partial N_i}\right) N_i F_i(N) \tag{4}$$

THEOREM 1 The positive equilibrium N^* of (3) is globally asymptotically stable if there exists a Liapunov function $V(N)$ such that $\dot{V}(N)$ is nonpositive in the positive orthant and it does not vanish identically along a nontrivial solution of (3) except for $N(t) = N^*$.

This theorem follows immediately from LaSalle's extension of the direct method of Liapunov, (see [4]).

Let $V_i(N_i)$ be a Liapunov function for a stable single species model whose population is N_i. If (3) describes a multispecies system in which each species is self-regulating and the interspecific interactions are relatively weak, then a candidate function is

$$V(N) = \sum_{i=1}^{m} c_i V_i(N_i) \tag{5}$$

where $c_1, c_2, \ldots, c_m$ are positive constants. Usually $c_1, c_2, \ldots, c_m$ are chosen by trial and error so that $\dot{V}(N)$ is nonpositive in the positive orthant. If $V_i(N_i)$ has the form given in (2), then

$$V(N) = \sum_{i=1}^{m} c_i \int_{N_i^*}^{N_i} [h_i(s)/g_i(s)]ds \tag{6}$$

where $h_i(s)$ and $g_i(s)$ have the same properties as $h(s)$ and $g(s)$ of (2).

The simplest expressions for $h_i(s)$ and $g_i(s)$ are $h_i(s) = s - N_i^*$ and $g_i(s) = s$. Using these functions we deduce that a positive equilibrium of (3) at N^* is globally asymptotically stable if there exist positive constants $c_1, c_2, \ldots, c_m$ such that the function

$$\dot{V}(N) = \sum_{i=1}^{m} c_i (N_i - N_i^*) F_i(N) \tag{7}$$

is negative semidefinite in the positive orthant and $\dot{V}(N)$ does not vanish identically along a nontrivial solution of (3) except for $N = N^*$.

Suppose the general Lotka-Volterra model

$$\dot{N}_i = N_i\left[b_i + \sum_{s=1}^{m} a_{is}N_s\right], \qquad i = 1, 2, \ldots, m \tag{8}$$

has a positive equilibrium at N^*. Let $A = (a_{is})$ and $C = \mathrm{diag}(c_1, c_2, \ldots, c_m)$. Using (7) we deduce that the equilibrium N^* is globally asymptotically stable if there exists a positive diagonal matrix C such that $CA + A^TC$ is negative semidefinite and the function

$$\dot{V}(N) = (1/2)(N - N^*)^T(CA + A^TC)(N - N^*) \tag{9}$$

does not vanish identically along a nontrivial solution of (8).

EXAMPLE 2 Consider the linear food chain model

$$\begin{aligned} \dot{N}_1 &= N_1[b_1 - a_{12}N_2] \\ \dot{N}_2 &= N_2[-d_2 + e_2a_{12}N_1 - a_{23}N_3] \\ \dot{N}_3 &= N_3[-d_3 + e_3a_{23}N_2 - a_{33}N_3] \end{aligned} \tag{10}$$

where b_1, d_2, d_3, e_2, e_3, a_{12}, a_{23} and a_{33} are positive constants.

It has an equilibrium at N^* where $N_2^* = b_1/a_{12}$, $N_3^* = (e_3a_{23}N_2^* - d_3)/a_{33}$, and $N_1^* = (a_{23}N_3^* + d_2)/(e_2a_{12})$. Suppose N^* is positive. If $c_2 = 1/e_2$ and $c_3 = 1/(e_2e_3)$, then $\dot{V}(N)$ of (9) reduces to

$$\dot{V}(N) = -\left(\frac{a_{33}}{e_2e_3}\right)(N_3 - N_3^*)^2 \tag{11}$$

By Theorem 1 the equilibrium N^* is globally asymptotically stable.

EXAMPLE 3 The Leslie prey-predator model (see p. 84 in [5]) is

$$\dot{N}_1 = N_1\left[\frac{r(K - N_1)}{K} - \frac{aN_2}{b + N_1}\right]$$

$$\dot{N}_2 = uN_2\left[1 - \frac{wN_2}{N_1}\right] \tag{12}$$

where r, K, a, b, u and w are positive constants.

Suppose (12) has a positive equilibrium at N*. If $h_1(s) = s - N_1^*$, $g_1(s) = s^2$, $h_2(s) = s - N_2^*$ and $g_2(s) = s$, then

$$\dot{V}(N) = \left\{(N_1 - N_1^*)\left[\frac{r(K - N_1)}{K} - \frac{aN_2}{b + N_1}\right] + c_2u(N_2 - N_2^*)(N_1 - wN_2)\right\}/N_1 \tag{13}$$

Therefore the equilibrium N* is globally asymptotically stable if there exists a positive constant c_2 such that $\dot{V}(N)$ of (13) is negative definite in the positive quadrant of the (N_1, N_2)-space.

THE POINCARÉ TRANSFORMATION

Let $\dot{V}(N)$ be the function in (4). Geometrically the condition that $\dot{V}(N)$ is negative semidefinite in the positive orthant implies that the angle between the velocity vector $(\dot{N}_i)$ and the normal of a level surface of V(N) is obtuse or is a right angle at every point in the positive orthant. Under a change of variables the direction of the velocity vector at a point may be altered. It follows that a preliminary change of variables may make it easier to apply the direct method of Liapunov to a given model. Hsu [3] has shown that the Poincaré transformation is a useful change of variables for some Lotka-Volterra models. It will be interesting to discover other useful classes of transformations.

We shall examine how the Poincaré transformation can be applied to a class of nonlinear population models. From (7) we deduce that it is desirable for $\partial F_1(N^*)/\partial N_1$ to be nonpositive. If it is positive, it may be worthwhile to subject (3) to the Poincaré transformation

$$N_1 = \frac{1}{x_1}, \; N_s = \frac{x_s}{x_1}, \qquad s = 2, 3, \ldots, m \tag{14}$$

We get

$$\begin{aligned} \dot{x}_1 &= -x_1 F_1(1/x_1, x_2/x_1, \ldots, x_m/x_1) \\ \dot{x}_s &= x_s(F_s - F_1) \end{aligned} \tag{15}$$

This transformation maps N^* into $x^* = (1/N_1^*, N_2^*/N_1^*, \ldots, N_m^*/N_1^*)$.

Using Theorem 1 and (7) we deduce that x^* is globally asymptotically stable if there exists positive constants $c_1, c_2, \ldots, c_m$ such that the function

$$\dot{V}(x) = -c_1(x_1 - x_1^*)F_1 + \sum_{s=2}^{m} c_s(x_s - x_s^*)(F_s - F_1) \tag{16}$$

is negative semidefinite in the positive orthant and $\dot{V}(x)$ does not vanish identically along of nontrivial solution of (15).

If $\dot{V}(x)$ is not negative semidefinite in the positive orthant, it may be worthwhile to subject model (15) to another Poincaré transformation.

EXAMPLE 4 Consider the prey-predator model

$$\begin{aligned} \dot{N}_1 &= N_1(0.25 + N_1 - 5N_2) \\ \dot{N}_2 &= N_2(-0.5 + 2N_1 - N_2) \end{aligned} \tag{17}$$

It has a locally stable equilibrium at $N^* = (1.0, 0.25)$.

We note that the N_1 population is not self-regulating. Therefore we apply the Poincaré transformation, $N_1 = 1/x_1$ and $N_2 = x_2/x_1$. We get

$$\begin{aligned} \dot{x}_1 &= x_1(-1 - 0.25x_1 + 5x_2)/x_1 \\ \dot{x}_2 &= x_2(1 - 0.75x_1 - x_2)/x_1 \end{aligned} \tag{18}$$

We note that each component in the transformed model is self-regulating.

If $c_1 = 0.75$ and $c_2 = 5$ then (16) gives

$$\dot{V}(x) = -[0.1875(x_1 - 1)^2 + 5(x_2 - 0.25)^2]/x_1 \tag{19}$$

Clearly $\dot{V}(x)$ is negative definite in the positive orthant. Therefore x^* of (18) is globally asymptotically stable. This implies that the equilibrium N^* of (17) is globally asymptotically stable.

ACKNOWLEDGMENT

This research partially supported by the National Research Council of Canada (Grant number A-3990).

REFERENCES

1. B. S. Goh, "Robust stability concepts for ecosystem models," *Theoretical Systems Ecology*, E. Halfon (Ed.), Academic Press, New York, Chap. 19, 467-487.

2. S. B. Hsu, "On global stability of a predator-prey system," *Math. Biosc.* 39(1978), 1-10.

3. S. B. Hsu, "The application of the Poincaré transform to the Lotka-Volterra model," *J. Math. Biol.* 6(1978), 67-73.

4. J. P. LaSalle and S. Lefschetz, *Stability by Liapunov's direct method with applications*, Academic Press, New York (1961).

5. R. M. May, *Stability and complexity in model ecosystems*, 2nd ed., Princeton University Press, Princeton, New Jersey (1974).

POPULATION DYNAMICS IN PATCHY ENVIRONMENTS

Alan Hastings

Department of Pure and Applied Mathematics
Washington State University
Pullman, Washington

INTRODUCTION

Ever since Gause's [3] pioneering studies, it has been clear to biologists that real biological predator prey systems in simple environments are highly unstable. The importance of spatial heterogeneity as a stabilizing factor was emphasized by Huffaker's [9] experimental work with mites and the natural system of the cactus moth, *Cactoblastis cactorum*, and its host, the prickly pear, *Opuntia Sp.* [2].

A related idea is that of predator mediated coexistence of prey and the stabilizing effect of disturbance in a spatially heterogeneous system emphasized by Paine [12] and others (see references in [1, 6]). The

Current affiliation:
Department of Mathematics, University of California, Davis, California.

formulation of models appropriate for understanding these systems has been an active area of research (e.g., [1, 5, 6, 7, 8, 10, 11] and refs. therein).

In this paper I will review and develop some models appropriate for studying the dynamics of interacting species occurring in an environment consisting of a large number of discrete, identical, patches. The results presented here are derived elsewhere ([5, 6, 7, 8]).

ONE PREDATOR-ONE PREY MODELS

The first step is to discuss those factors that will be included in the model. The environment will consist of a large number of identical patches, so a deterministic model based on fractions of patches in various states will be developed. The state of a patch will be determined only by which species are present and how long they have been present in that particular patch. Population levels will be ignored--this is equivalent to assuming that changes in population levels take place on a faster time scale than the other processes being modelled.

Every patch in the model will be assumed equally accessible from every other patch. Thus long range dispersal is assumed to be the predominant means of population exchange and short range dispersal is ignored. The rate at which a species colonizes suitable patches is assumed to depend linearly on the number of patches it occupies, and in the simple models presented here dispersal will take place on a faster time scale than that of the other parameters in the model. Given these assumptions, a very general model can be written. As a first example consider a one predator one prey system that obeys the following specific dynamics of [7].

1. An *empty* patch can be invaded only by *prey*, creating a *prey* patch.
2. The patch remains in this state until invaded by the predator, becoming then a *predator* patch.
3. *Predator* patches are assumed to return to the *empty* state at a constant rate.
4. For simplicity, it is assumed that prey disperse only from *prey* patches; hence they invade *empty* patches at a rate proportional to the fraction of patches that are *prey*.
5. Predators invade *prey* patches at a rate proportional to the fraction of *predator* patches.

Let P_A be the fraction of prey patches, P_c the fraction of predator patches, D_A and D_c the constants determining the colonization rates of prey and predator, respectively. The decay rate of predator patches is normalized to be one. Then the following equations result:

$$\frac{dP_A}{dt} = D_A P_A (1 - A_A - P_c) - D_c P_A P_c$$
$$\frac{dP_c}{dt} = D_c P_A P_c - P_c \tag{1}$$

Clearly, the variables P_A and P_c are constrained to the *feasible* region

$$0 \leq P_A \leq 1$$
$$0 \leq P_c \leq 1 - P_A \tag{2}$$

As shown in Hastings [5], the asymptotic behavior of the system (1) is completely determined by the magnitude of D_c. If $D_c > 1$ there is a globally (among positive *feasible* initial conditions) stable, feasible, nontrivial equilibrium point. If $D_c < 1$, the equilibrium with $P_A = 1$, $P_c = 0$ is globally stable. Thus the predators dispersal ability completely determines the long range behavior.

The assumption that the prey patches return to the empty state at a constant rate is unrealistic and will be altered here. This assumption will be strictly true only in the probability that a *predator* patch returns to the empty state is independent of the time since it was created. Given that the predator-prey dynamics are unstable, this is unlikely to be true. Instead, I will consider a modification used in [5, 6, 7]. Here, the *predator* patches all last a fixed time and then return to the empty state--the probability of return to the empty state is totally determined by the "age" of the patch. Without loss of generality, the "lifetime" of the predator patches can be normalized to be 1, so the populations will obey the equations:

$$\frac{dP_A}{dt} = D_A (1 - P_A - P_c) P_A - D_A P_A P_c$$
$$P_c(t) = D_c \int_{t-1}^{t} P_A(s) P_c(s) ds \tag{3}$$

When initial conditions are specified on an interval, a well-posed problem results [6]. Again, if $D_c < 1$, the equilibrium $P_A = 1$, $P_c = 0$ is globally stable [6]. If $D_c > 1$, there exists a nontrivial feasible equilibrium point, whose local stability was determined in [6]. The general feature is that if D_c is small enough the equilibrium is locally stable, and if D_c is too large it is unstable. As indicated in [7], when stability is lost, a Hopf bifurcation occurs and periodic orbits should exist.

TWO PREY-ONE PREDATOR MODELS

Here a model appropriate for studying the phenomenon of predator mediated coexistence will be considered. Although a more complex model based on the "age dependent" model presented above has been studied [6] here I will consider only a simpler model [8].

The model will have four possible states for patches: E, empty; L, the inferior competitor (prey) alone; W, the competitive dominant (prey) alone; C, predator.

The rules governing the dynamics within the patches are:

1. An E patch can be invaded by L creating the state L, or by W creating W.
2. Species W invades L patches at the same rate as empty patches, creating W patches. Species L cannot invade W patches.
3. The predator invades only L or W patches, creating C patches.
4. At a fixed rate (set equal to one without loss of generality) C patches return to state E.
5. The colonization rates of the prey species are proportional to the number of patches they occupy.
6. The colonization rate of the generalist predator will be proportional to the number of C patches, and will be the same for L or W patches.

Under these assumptions, it is clear that in the absence of the predator, L will be eliminated from the system, every patch ending up as WW. Thus, coexistence, if it occurs, must be an effect of the predator. It will also be shown that unless L has the advantage of a higher mobility than W or negative predator preference, L will be eliminated.

Let P_L be the fraction of L patches, P_W the fraction of W patches, P_c the fraction of C patches, $P_E = 1 - P_L - P_W - P_c$ the fraction of empty patches. Again attention is restricted to the *feasible* region where all of these variables are nonnegative. Then if D_L, D_W, and D_c are the constant measures of colonization abilities, it is clear that the equations of the model are:

$$\frac{dP_L}{dt} = D_L P_L P_E - D_W P_L P_W - D_c P_L P_c$$

$$\frac{dP_W}{dt} = D_W P_W(1 - P_W - P_c) - D_c P_W P_c \qquad (4)$$

$$\frac{dP_c}{dt} = D_c P_c(P_L + P_W) - P_c$$

Since the predator will survive only if $D_c > 1$, attention here is restricted to that case. The asymptotic behavior of the model is determined by the inequalities [8]

$$D_L > D_W\left(\frac{D_c^2 + D_W}{D_c^2 - D_c}\right) \qquad (5)$$

and

$$D_L < D_W\left(\frac{D_c^2 + D_L}{D_c^2 - D_c}\right) \qquad (6)$$

If (5) and (6) hold, then there is a feasible, nontrivial equilibrium with all three species present, approached by all feasible, positive initial conditions. This is shown using techniques like those in Goh [4].

If (6) holds, but (5) is violated L will be eliminated and if (5) holds, but (6) is violated W will be eliminated.

It has been conjectured, and there is some data, that intermediate levels of predation lead to the highest diversity of prey (e.g., Paine and Vadas [13] and [6]). Some limiting cases of this simple model agree with this. If $D_c \to \infty$, which is equivalent to very high predation (5) becomes

$$D_L > D_W$$

and (6) becomes

$$D_L < D_W$$

so coexistence of prey is impossible. If D_c is close to one (low predation) (5) is satisfied only for very large values of D_L, so again coexistence is less likely. A further discussion and extensions of this model are found elsewhere [1, 6, 8].

REFERENCES

1. H. Caswell, "Predator mediated coexistence: a nonequilibrium model," *Amer. Nat.* 112(1978), 127-154.

2. A. P. Dodd, "The biological control of prickly pear in Australia," in *Biogeography and Ecology in Australia*, A. Keast, R. L. Crocker, and C. S. Christian (Eds.). Monogr. Biol. 8(1959), 565-577.

3. G. F. Gause, *The Struggle for Existence*. Williams and Williams, Baltimore (1934).

4. B. S. Goh, "Global stability in many-species systems," *Amer. Nat.* 111 (1977), 135-143.

5. A. Hastings, "Spatial heterogeneity and the stability of predator prey systems," *Theo. Pop. Bio.* 12(1977), 37-48.

6. A. Hastings, "Spatial heterogeneity and the stability of predator prey systems: predator-mediated coexistence," *Theo. Pop. Bio.* 14(1977), 380-395.

7. A. Hastings, "Spatial heterogeneity and the stability of predator prey systems: population cycles," pp. 607-618 in *Applied Nonlinear Analysis*. V. Laksmikantham (Ed.), Academic Press, New York (1979).

8. A. Hastings, "The effects of disturbance and predation on competitive systems," in preparation.

9. C. B. Huffaker, "Experimental studies on predation: dispersion factors and predator-prey oscillations," *Hilgardia* 27(1958), 343-383.

10. S. A. Levin, "Spatial patterning and the structure of ecological communities," in *Some Mathematical Questions in Biology*, VII. S. A. Levin (Ed.), (Lectures on Mathematics in the Life Sciences, Vol. 8, American Mathematical Society) Providence, Rhode Island (1976).

11. S. A. Levin, "Population dynamics in heterogeneous environments," *Ann. Rev. Ecol. Systematics* 7(1976), 287-310.

12. R. T. Paine, "Food web complexity and species diversity," *Amer. Nat.* 100(1966), 65-75.

13. R. T. Paine and R. L. Vadas, "The effects of grazing by sea urchins," *Strobgylocentrotus* (spp., on benthic algal populations) *Limnol. Oceanog.* 14(1969), 710-719.

LIMIT CYCLES IN A MODEL OF B-CELL STIMULATION

Stephen J. Merrill

Department of Mathematics and Statistics
Marquette University
Milwaukee, Wisconsin

INTRODUCTION

The B-cell is a variety of lymphocyte which is responsible for the production of proteins called antibody which protect us from invasion by substances which our body finds to be foreign. These foreign substances are called antigens.

This "humoral" response is one of the two broad responses of our immune system (the other is named the "cell-mediated" response). The humoral response is a natural target for modeling because of the readily identifiable and measurable cell product (antibody) which is produced by one cell type, the B-cell. Research has shown however that for most antigens, other cells, e.g., T-cells and macrophages, must in some way process the antigen before the B-cell is ready to act. The antigen discussed here will be any which is "T-independent" (not requiring this processing).

In models of the production of antibody when the lymphocyte senses the presence of antigen, several things must be considered:

1. The stimulation of the B-cell--that is, the awakening of the B-cell to the presence of antigen--is a threshold phenomenon. Small amounts of antigen may cause the production of little or no antibody while larger amounts may turn our lymph nodes and spleen into veritable antibody factories.
2. The antibody produced by the B-cells floats around looking for antigen. When it finds the antigen, it attaches and either triggers a sequence of enzymes (an enzyme cascade) called the "complement system," which is effective against cellular antigen, or "opsonizes" the antigen, making it tasty for the scavenger cells--the macrophage.
3. The stimulation of the B-cell occurs on a much faster time scale than the response itself (minutes as opposed to days). The fact of two time scales will be incorporated into the model.

A standard reference for all aspects of the immune response is [6].

In this model of the stimulation of B-cells and the corresponding production of antibody, assumptions are made about both the mechanism of the stimulation as dictated by the form of the threshold and the dynamics of the interaction of antibody and antigen.

CONSTRUCTION OF THE MODEL

In the population of B-cells, for a given antigen, only relatively few will be able to respond (with stimulation) for that particular antigen. We will limit the discussion to the fraction of B-cells able to respond to our fixed antigen.

Let

Y = number of B-cells which could respond to the antigen, but as yet have not.

Z = number of B-cells stimulated by the presence of the antigen.

Then $x = (Y - Z)/(Y + Z)$ is a dimensionless number between -1 and 1 which measures the amount of stimulation in the system. x is 1 when all B-cells are unstimulated and x is -1 when all are stimulated.

The assumption of the form of the threshold is in two parts:

1. There is a critical value (threshold value) of antigen b^o so that if the antigen dose is larger than b^o, x will be near -1. If the dose is less than b^o, x will be near 1.

2. If antigen is removed rapidly from the system ("washed out" if *in vitro*), the stimulated cells would quickly revert to an unstimulated state. That is, antigen is required at all times between stimulation and production of antibody. It is further assumed that this "shut-off point" happens at a value of antigen b_o less than b^o. (It takes less to keep the system going than to start it.)

These assumptions lead to Figure 10.1.

This type of threshold can be simulated by the following: Consider the following curve shown in Figure 10.2:

$$x^3 - \frac{1}{2}x + b - \frac{1}{2} = 0$$

If the solutions of our model are required to stay on the curve above, then when b is greater than b^o, the solution must be near x = -1, and when b is less than b_o, the solution must be near x = +1. If $b_o < b < b^o$, the state of the B-cells remains as it was.

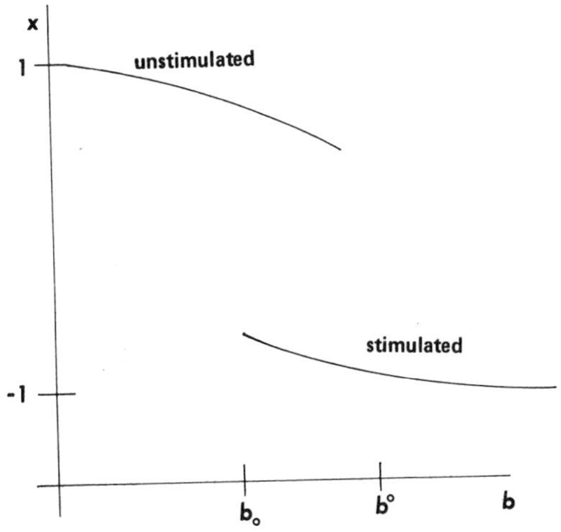

Fig. 10.1. Stimulation (x) as depending on antigen dose.

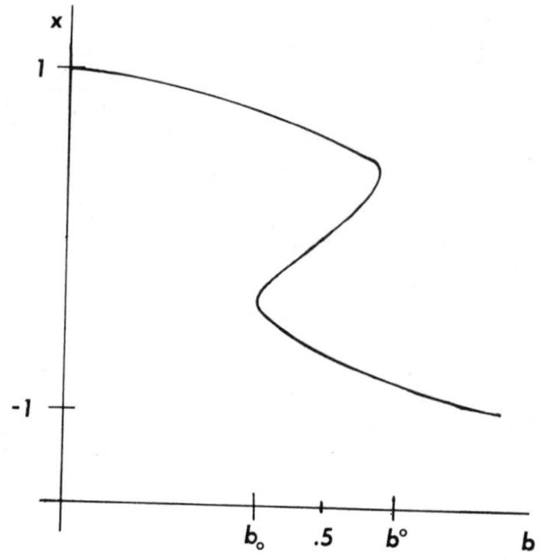

Fig. 10.2. The curve $x^3 - \frac{1}{2}x + b - \frac{1}{2} = 0$.

If there is antibody in the system at the time of the introduction of antigen, the response is smaller--that is, fewer B-cells will be stimulated due to competition for antigen. One way to incorporate this in the model is to parameterize the curve so that the threshold nature changes as the amount of antibody, a, in the system is varied.

The following surface (see Figure 10.3) results from parameterizing the curve by $x^3 + (a - \frac{1}{2})x + b - \frac{1}{2} = 0$.

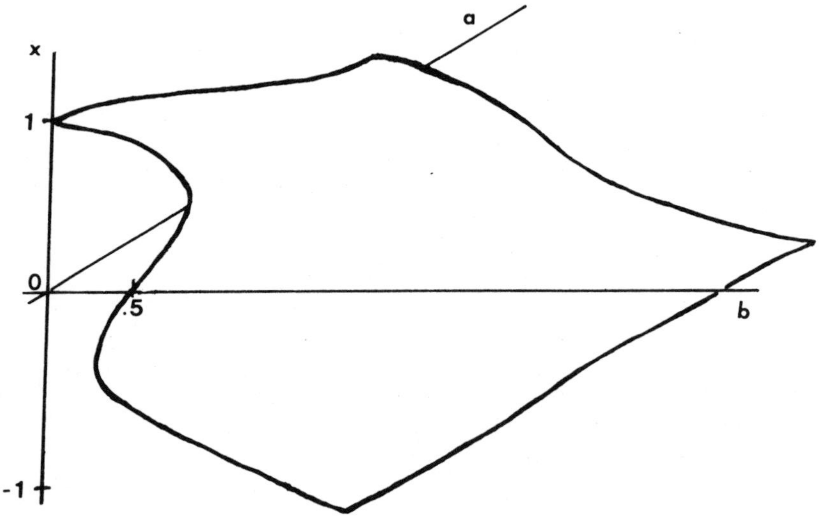

Fig. 10.3. The surface $x^3 + (a - \frac{1}{2})x + b - \frac{1}{2} = 0$.

The threshold gets less pronounced as a increases. (This surface is the cusp singularity of Thom [18].)

The path of stimulation is not instantaneous, but it is very rapid. The solution of the model equations will be required to stay on or near this surface except for very rapid transit movements from x near 1 to x near -1, or vice versa. This can be done mathematically by the following system:

$$\varepsilon\dot{x} = -(x^3 + (a - \tfrac{1}{2})x + b - \tfrac{1}{2})$$

$$\left.\begin{array}{l}\dot{a} = \\ \\ \dot{b} = \end{array}\right\} \text{ equations for dynamics of antibody and antigen}$$

where $1 >> \varepsilon > 0$.

This singularly perturbed system has the property that solutions will change the x coordinate rapidly if off the surface. Note: In enzyme kinetic equations (Briggs-Haldane) the ε is assumed to be 0 and an algebraic equation results--that is, the very rapid effects are usually ignored.

The equations for antibody a and antigen b (with the time scale 1 time unit = half life of the antibody class considered roughly 20 days) are

$$\dot{a} = \begin{matrix}\text{amplification}\\ \text{constant}\end{matrix} \cdot \begin{pmatrix}\text{amount of}\\ \text{stimulation}\end{pmatrix} - \begin{matrix}\text{decay of}\\ \text{antibody}\end{matrix} - \begin{pmatrix}\text{mass action}\\ \text{interaction}\\ \text{of a and b}\end{pmatrix}$$

$$= \delta\left(\frac{1 - x}{2}\right) - a - \gamma_1 ab$$

and

$$\dot{b} = -\gamma_1 ab + \gamma_2 b$$

Exponential growth of antigen is allowed as no crowding effects would take effect in the presence of an adequate immune defense.

The model is

$$\varepsilon\dot{x} = -(x^3 + (a - \tfrac{1}{2})x + b - \tfrac{1}{2})$$

$$\dot{a} = \tfrac{\delta}{2}(1 - x) - a - \gamma_1 ab \qquad (1)$$

$$\dot{b} = -\gamma_1 ab + \gamma_2 b$$

$$-1 \le x(0) \le 1, \quad a(0) \ge 0, \quad 0 \le b(0), \quad 1 >> \varepsilon > 0, \quad \delta > 0, \quad \gamma_i > 0$$

BEHAVIOR OF THE MODEL SOLUTIONS

If $\gamma_2 < 0$ (antigen does not have the ability to grow), all solutions of (1) asymptotically converge to the equilibrium point $x = 1$, $a = 0$, $b = 0$ ([10]). Typical solutions are shown in Figure 10.4. These show the unresponsiveness to subthreshold stimulation.

If $\gamma_2 > 0$ (the antigen can replicate), the mathematical setting is more interesting. In [11], periodic solutions of two types were shown to exist to (1). Also, in [11], a positively invariant compact region was displayed when δ is large enough and ε is sufficiently small. For one of those types of periodic solutions it is possible to show that it is a limit cycle.

EXISTENCE OF LIMIT CYCLES ($\gamma_2 > 0$)

Consider the "constrained" system:

$$0 = x^3 + (a - \tfrac{1}{2})x + b - \tfrac{1}{2} = g(x, a, b)$$

$$\dot{a} = \tfrac{\delta}{2}(1 - x) - a - \gamma_1 ab \qquad (2)$$

$$b = -\gamma_1 ab + \gamma_2 b$$

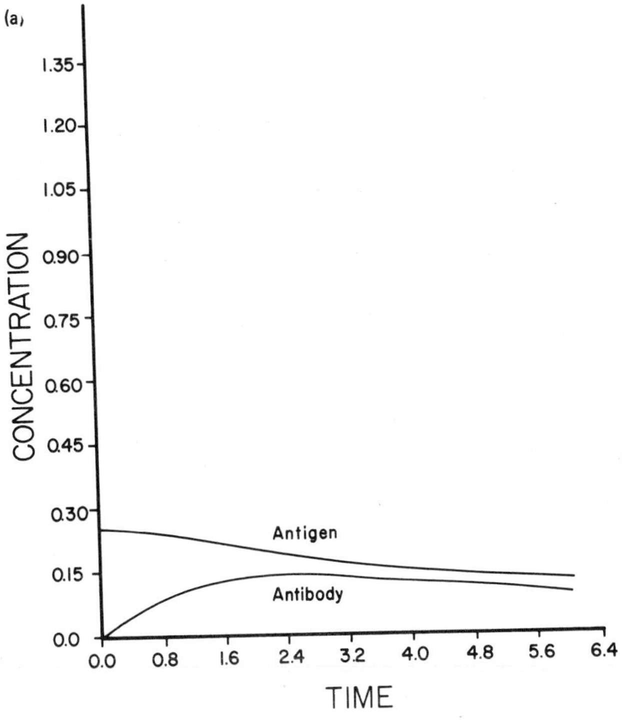

Fig. 10.4a. Low-dose unresponsiveness.

In this system, solutions must stay on the manifold given by $0 = x^3 + (a - \frac{1}{2})x + b - \frac{1}{2}$. Solutions in general fail to be continuable even if they stay within a compact set unless they stay away from the "catastrophe set" given by $\{(x, a, b) | \partial g(x, a, b)/\partial x = 0\}$. Generalization of the idea of a solution is possible. For instance, [8, 16, 17] which gets around this difficulty. Anosov [1] and Chang [4] have shown that, if (2) has a periodic solution of a particular type, then (1) will also have a periodic solution near that of (2) if ε is sufficiently small.

Periodic Solutions to (2)

It is now shown that the degenerate system (2) has periodic solutions which arise from a Hopf bifurcation on the manifold $g(x, a, b) = 0$. Consider

$$\dot{X} = F(X, \delta) \tag{3}$$

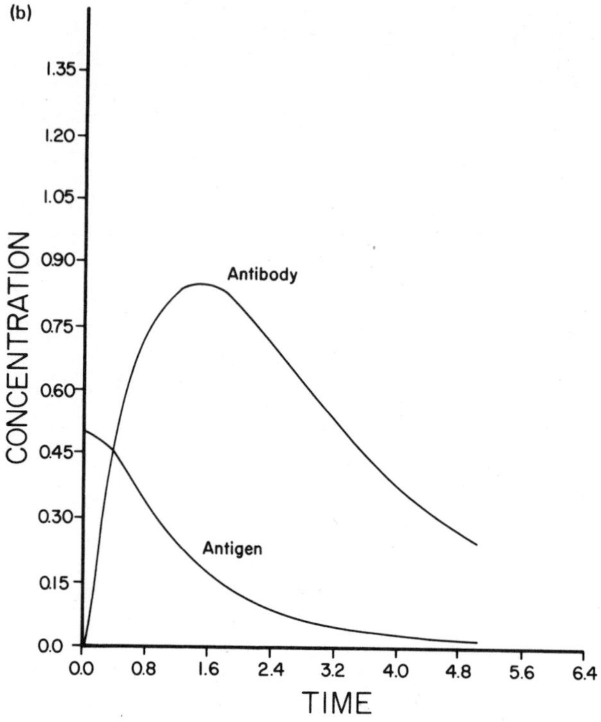

Fig. 10.4b. Typical response to a threshold dose of antigen.

where δ is a real scalar parameter. Assume there is a function $a(\delta)$ defined for δ in some interval such that

$$F(a(\delta), \delta) = 0$$

Let

$$A(\delta) = F_x(a(\delta), \delta)$$

(the Jacobian Matrix).

THEOREM (Hopf [9]) Let $F(x, \delta)$ be analytic in x and δ in a neighborhood of $(a(\delta_c), \delta_c)$. Let the eigenvalues of $A(\delta)$ for δ sufficiently near δ_c be $u(\delta) + iv(\delta)$. Then, if (i) $u(\delta_c) = 0$, $v(\delta_c) \neq 0$ and (ii) $u_\delta(\delta_c) \neq 0$ holds, there will be a bifurcation of periodic solutions of (3) from $\delta = \delta_c$, either for $\delta > \delta_c$, $\delta < \delta_c$ or $\delta = \delta_c$.

In [11] it was shown that when

$$\delta_c = 2\,\frac{(\gamma_2/\gamma_1) + \gamma_2 b_o}{1 - x_o}$$

where $(x_o, \gamma_2/\gamma_1, b_o)$ lies on the manifold and satisfies

$$\frac{\gamma_2}{\gamma_1} = \frac{(3x^2 - \frac{1}{2})(1 - x)}{2x - 1} \tag{4}$$

that the Hopf theorem is satisfied and periodic solutions to (2) exist.

In [11] using the stability formula from [9, p. 126], it was shown that the periodic solutions are stable and occur for $\delta < \delta_c$ if

$$\frac{\delta_c}{2}\left(-3x_{\delta_c}^2 + \frac{\gamma_2}{\gamma_1} - \frac{1}{2}\right) + \gamma_1\left(3x_{\delta_c}^2 + \frac{\gamma_2}{\gamma_1} - \frac{1}{2}\right)^3 > 0 \tag{5}$$

In Figure 10.5, the curve (4) is shown in the a-b plane along with the approximate region where (5) is satisfied.

The Theorem of Chang

THEOREM (Chang [4, Theorem 1]) Consider the system

$$\begin{aligned} \dot{y} &= f(y, z, \varepsilon) \\ \varepsilon\dot{z} &= g(y, z, \varepsilon) \end{aligned} \tag{6}$$

where $\varepsilon > 0$, $y \in \mathbb{R}^m$, $z \in \mathbb{R}^n$ and its associated degenerate system

$$\begin{aligned} \dot{y} &= f(y, z, 0) \\ 0 &= g(y, z, 0) \end{aligned} \tag*{$(6)_o$}$$

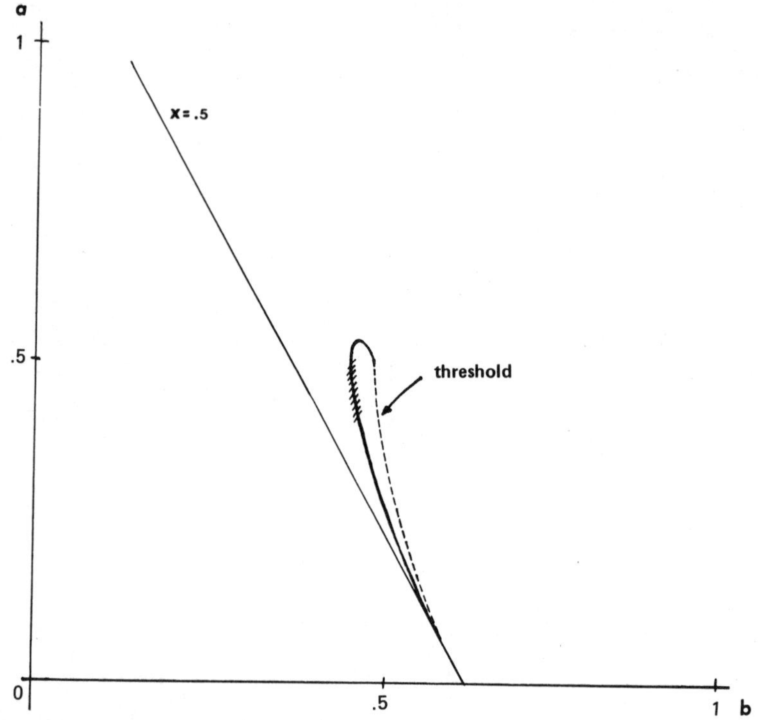

Fig. 10.5. The curve (4). Hatched area of the curve is a typical region satisfying (5). This region actually depends on γ.

Suppose

(I) the equation $g(y, z, 0) = 0$ has a continuously differentiable solution $z = \phi(y)$ for $y \in D$, some region in y-space.

(II) the degenerate system $(6)_0$ has in D a nonconstant periodic solution u(t) with the period T in t.

(III) the variational equation of $(6)_0$

$$\dot{\nu} = f_y[u(t)]\nu$$

admits a single multiplier equal to 1.

(IV) every eigenvalue of $g_z(u(t), \phi(u(t)), 0)$ has real part different from zero.

Then for ε sufficiently small, (6) has a periodic solution $y_\varepsilon(t)$, $z_\varepsilon(t)$ with period $T + o(1)$ in t such that as $\varepsilon \to 0$, the closed path of this solution tends to the closed path of the periodic solution $(u(t), \phi(u(t))$ of the degenerate system.

Furthermore, there exists $p > 0$ such that (6) has no other periodic solution (apart from translations in t) whose path remains in the p-neighborhood of $(u(t), \phi(u(t)))$ and whose period differs from T by less than p.

The conditions (I, II, and IV) in the case of (1) and (2) are satisfied if the periodic solution arising from the bifurcation stays away from the catastrophe set (which it does for $\delta_c - \delta$ small).

Hypothesis (III) requires some information on how the stability was computed and Floquet theory.

Recall [7] if P(t) is a periodic matrix of period p (all entries periodic), then the fundamental matrix Y of

$$\dot{y} = P(t)y \tag{7}$$

has the form

$$Y(t) = Z(t)e^{Rt}$$

where Z(t) is periodic with period p and the eigenvalues of e^{Rt} are the (characteristic roots or) multipliers while the eigenvalues of $R(\text{mod}(2\Pi i)/p)$ are the characteristic exponents.

If u(t) is a non-constant periodic solution of $\dot{x} = F(x)$, then the variational equation

$$\dot{y} = p(t)y \tag{8}$$

where $p(t) = F_x[u(t)]$ will always have one multiplier equal to 1 (one exponent $= 0 \ (\text{mod}(2\Pi i/p))$ as $\dot{u}(t + p) = \dot{u}(t)$ and $\dot{u} = F[u(t)]$ satisfies (8).

So the stability of the periodic solution u(t) depends on the sign of the nonzero characteristic exponent. The sign of this exponent is indirectly calculated when using the stability formula [9, p. 126]. It has been shown [9, p. 204-5] that it agrees in sign with the value V''' in the stability calculation. Thus we have in the case of Equations (1) and (2), condition (III) holding for the bifurcating solution. Moreover, since the variational equation of (2) about the periodic solution has only one multiplier with modulus 1 when Equation (5) holds, Theorem 4 of Chang [4] implies that all solutions near the periodic solution of (1) either leave at an exponential rate or approach the periodic solution.

Construction of a Torus Containing a Periodic Solution of (1)

Consider one of the stable periodic solutions of (2). This solution lies on the surface $g(x, a, b) = 0$. As it is attracting, there is an annulus about it so that all solutions of (2) in the annulus approach the periodic solution and the vector field is transversal to the boundary of the annulus.

Extend the boundaries "vertically" (in the x and negative x direction) and consider the vector field determined by (1).

As the vector field of (1) coincides with that of (2) on the surface away from places where $\partial g/\partial x = 0$, for sufficiently small vertical perturbations of the annulus boundaries, the vector field of (1) will point inward. Let d = inf {vertical displacements such that the perturbed annulus boundaries will be transverse to the flow of (1)} > 0. Extend the annulus vertically (up and down) by d.

Set

$$m_1 = \min g(x, a, b) > 0$$

where the minimum is taken over the upper edges of the perturbed annulus and set

$$m_2 = \max g(x, a, b) < 0$$

where the maximum is taken over the lower edges.

Bound the perturbed annulus by

$$\{(x, a, b) \mid g(x, a, b) = \frac{m_1}{2}\} \quad \text{and} \quad \{(x, a, b) \mid g(x, a, b) = \frac{m_2}{2}\}$$

The resulting torus (see Figure 10.6) is positively invariant and contains a periodic solution of (1) for all ε sufficiently small. (Such a torus can easily be made within p radius of our periodic solution of (2).) By Chang's Theorem 4, since no solution of (1) can leave this region, all solutions starting within the torus must approach the periodic solution of (1) which is thus a limit cycle.

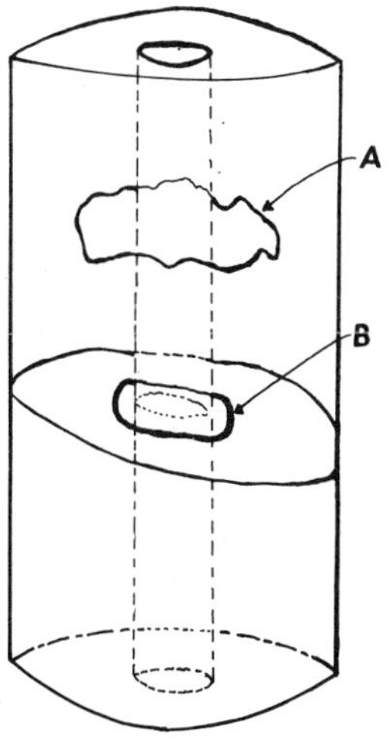

Fig. 10.6. Torus as constructed containing periodic solutions of (1), labeled A, and (2), labeled B, for small ε.

BIOLOGICAL MEANING OF A STABLE PERIODIC SOLUTION

The existence of a stable periodic solution in the model corresponds to a recurring infection. An excellent example may be found in the disease--Brucellosis--where the recurrent fever is called "undulant fever."

REFERENCES

1. D. V. Anosov, "On limit cycles in systems of differential equations with a small parameter in the highest derivatives," *Am. Math. Soc. Translations ser.* 2, 33:233-275.

2. G. I. Bell, "Predator-prey equations simulating an immune response," *Math. Biosci.* 16(1973), 291-314.

3. G. I. Bell, Alan S. Perelson and G. H. Pimbley, Jr. (Eds.), *Theoretical Immunology, Immunology Series* Vol. 8, Marcel Dekker, New York (1978).

4. K. W. Chang, "Two problems in singular perturbations of differential equations," *J. Aust. Math. Soc.* X(1969), 33-50.

5. J. Cronin, "Some mathematics of biological oscillations," *SIAM Rev.* 19(1977), 100-138.

6. H. N. Eisen, *Immunology* (reprinted from Davis, Dulbecco, Eisen, Ginsberg and Wood, *Microbiology*, 2nd ed.). Harper and Row, New York (1974).

7. J. K. Hale, *Ordinary Differential Equations*. Wiley-Interscience, New York (1969).

8. N. Levinson, "Perturbations of discontinuous solutions of nonlinear systems of differential equations," *Acta Math.* 82(1950), 71-106.

9. J. E. Marsden and M. McCracken, *The Hopf Bifurcation and its Applications*. Springer-Verlag, New York (1976).

10. S. J. Merrill, "A geometrical study of B-cell stimulation and humoral immune response," in *Nonlinear Systems and Applications*. V. Lakshmikantham (Ed.), Academic Press, New York (1977), 611-630.

11. S. J. Merrill, "A model of the stimulation of B-cells by replicating antigen," in two parts, *Math. Biosci.* 41(1978), 125-141 and 143-155.

12. S. J. Merrill, "Mathematical models of humoral immune response," Chapter 2 of this volume.

13. G. H. Pimbley, Jr., "Periodic solutions of predator-prey equations simulating an immune response," *Math. Biosci.* 20(1974), 27-51 and 21(1974), 251-277.

14. G. H. Pimbley, Jr., "Periodic solutions of third order predator-prey equations simulating an immune response," *Arch. Rat. Mech. Anal.* 55 (1974), 93-123.

15. G. H. Pimbley, Jr., "Bifurcation behavior of periodic solutions for an immune response," *Arch. Rat. Mech. Anal.* 64(1977), 169-192.

16. R. E. Plant, "Crustacean cardiac pacemaker model: an analysis of the singular approximation," *Math. Biosci.* 36(1977), 149-171.

17. F. Takens, "Constrained equations: a study of implicit differential equations and their discontinuous solutions," in *Lecture Notes in Mathematics*, P. Hilton (Ed.), Vol. 525, Springer-Verlag, Berlin (1975), 143-234.

18. R. Thom, *Structural Stability and Morphogenesis*, Benjamin, Reading, Mass. (1975).

OPTIMAL AGE-SPECIFIC HARVESTING POLICY FOR A CONTINUOUS TIME-POPULATION MODEL

Chris Rorres

Department of Mathematics
Drexel University
Philadelphia, Pennsylvania

Wyman Fair

Department of Mathematics
Drexel University
Philadelphia, Pennsylvania

INTRODUCTION

The problem of harvesting a population divided into discrete age classes by removing individuals from selective age classes was first investigated by Lefkovitch [5, 6]. Later, several authors (Beddington and Taylor [1], Rorres and Fair [8], Doubleday [4], Rorres [7]) investigated the problem of determining the optimal sustainable yield obtainable by such age-specific harvesting. For discrete populations, the determination of the optimal sustainable yield was shown to reduce to a Linear Programming problem when the underlying growth of the population was governed by a linear matrix model.

In this paper, we investigate the analogous problem for a population governed by a continuous-time rather than a discrete-time linear model. The effects of harvesting such a continuous-time population were investigated by Brauer [2] and Sánchez [9], but not with the purpose of determining the optimal sustainable yield. Results in this direction can also be found in Clark's text [3].

In the model to be presented, the dynamics of the population to be harvested are controlled by the age specific mortality rate $\mu(a)$ and the age-specific birth rate $b(a)$, where a is the age of an individual. As is standard for such models, we consider only the females of the population. It is assumed that $\mu(a)$ and $b(a)$ are such that without harvesting the population would grow without bound. Age-specific harvesting at a rate of $h(a)$ females of age a per unit time is then introduced to bring the population to an equilibrium level. The problem is to determine the harvest rate $h(a)$ which will maximize the yield of the resulting sustainable harvest, taking into account the economic value of the females as a function of age and the amount of capital the harvester has to raise the population and/or to harvest it. We show that the determination of $h(a)$ leads to an optimal control problem. In the first section of this paper, we show that if no upper bound is imposed on the harvest rate $h(a)$, then the optimal harvesting policy is impulsive and bimodal; i.e., only at most two ages are harvested, with the older being harvest completely. This is analogous to the solution of the discrete-time problem. In the second section, we impose an upper bound on the harvest rate and give a qualitative description of the resulting optimal harvesting policy.

IMPULSIVE HARVESTING

To state the problem to be examined, we first define the following three functions:

$u(a)$ = *population density*; the density of females of age a in the equilibrium population.

$\mu(a)$ = *mortality rate*; the natural mortality rate of females of age a.

$h(a)$ = *harvest rate*; the density of females of age a harvested per unit time.

These three quantities are related by the following basic equilibrium equation:

$$u'(a) = -\mu(a)u(a) - h(a), \qquad a \geq 0 \tag{1}$$

The initial density u(0) is the density of the new-born females and is governed by the condition

$$u(0) = \int_0^A b(a)u(a)da \quad (2)$$

where

b(a) = *birth rate*; the density of new-born daughters produced by one female of age a

The upper limit of integration A in Equation (2) is the age at which u(a) first vanishes in the equilibrium population. We admit the possibility that $A = \infty$. Next, setting

y(a) = *yield function*; the economic value of a single harvested female of age a

the economic yield of the harvest, Yld, is given by

$$\text{Yld} = \int_0^A y(a)h(a)da \quad (3)$$

More precisely, Yld is a yield rate--so much money per unit time.

Finally, we need a constraint which limits the size of the unharvested population. Defining

c(a) = *cost function*; the cost per unit time of maintaining a single unharvested female of age a

the constraint we shall use assumes the form

$$\int_0^A c(a)u(a)da = 1 \quad (4)$$

We have chosen economic units so that the harvester is constrained to spend one economic unit per unit time to maintain the total equilibrium population.

It is possible to give c(a) and Equation (4) alternate economic or ecological interpretations, but the above interpretation will suffice for our purposes.

The precise problem is then to find a harvest rate $h(a) \geq 0$ and terminal time A so that the yield Yld is maximized subject to Equations (1), (2), and (4) and the auxilliary conditions $u(a) > 0$ for $0 \leq a < A$ and $u(A) = 0$.

In order to restate this problem as a standard optimal control problem, we define the auxilliary variables

$$v(a) \triangleq \int_A^a b(\xi)u(\xi)d\xi \tag{5}$$

$$w(a) \triangleq \int_0^a c(\xi)u(\xi)d\xi \tag{6}$$

The problem then becomes:

PROBLEM 1 Given y(a), μ(a), b(a), and c(a), find a harvest rate h(a) and terminal time A which maximizes

$$\text{Yld} = \int_0^A y(a)h(a)da$$

subject to

$$u'(a) = -\mu(a)u(a) - h(a)$$

$$v'(a) = b(a)u(a)$$

$$w'(a) = c(a)u(a)$$

and

$$u(0) = -v(0), \qquad u(A) = 0$$

$$w(0) = 0, \qquad v(A) = 0$$

$$w(A) = 1$$

$u(a) \geq 0$

$h(a) \geq 0$

Since this is a linear control problem, the optimal policy is a bang-bang control; for each age either h(a) has its minimum value of zero (no harvesting of that age) or its maximum value (maximum harvesting of that age). We shall, however, initially allow h(a) to be unbounded. This will result in the optimal policy being attained by *impulsive* harvesting; that is, harvesting for which h(a) is a linear combination of Dirac delta functions. Consequently, harvesting occurs only at discrete ages at which the population density u(a) suffers jump discontinuities.

We next show that the optimal impulsive policy is *bimodal*--one in which only two age classes are harvested, the older class being harvested completely. We show this by converting the impulsive harvesting problem to a Linear Programming problem and then using previously known results. Suppose for some arbitrary impulsive harvesting policy the ages harvested are a_1, $a_2, \ldots, a_n$ where $0 = a_1 < a_2 < \ldots < a_n$. In order to simplify the formulas which follow, we adopt the following conventions: We always set $a_1 = 0$ even though no harvesting need occur at that age. If the population is harvested completely at the last harvested age a_n, we set $a_n = a_{n+1} = A$. If some females survive beyond that age, we set $a_{n+1} = A$ where as before A is the age at which u(a) first vanishes. The problem now is to determine how many females should be harvested at each of these n ages in order to maximize the yield.

Between harvesting ages, Equation (1) is simply $u'(a) = -\mu(a)u(a)$, and so u(a) is of the form $k_i\pi(a)$ between a_i and a_{i+1} where

$$\pi(a) \triangleq \exp\left(-\int_0^a \mu(\xi)d\xi\right) \tag{7}$$

The function $\pi(a)$ has the demographic interpretation of the probability of a female surviving to age a if she is not harvested. The constants k_i are determined by the values of $u(a_i+)$ as follows:

$$u(a) = u(a_i+)\pi(a)/\pi(a_i), \qquad a_i < a < a_{i+1},\ i = 1, 2, \ldots, n \tag{8}$$

Equation (2) can now be written as

$$u(a_1-) = u(0) = \int_0^A b(a)u(a)da$$

$$= \sum_{i=1}^{n} \int_{a_i}^{a_{i+1}} b(a)u(a_i+)\frac{\pi(a)}{\pi(a_i)}\,da = \sum_{i=1}^{n} A_i u(a_i+) \tag{9}$$

where we define A_i by

$$A_i \triangleq \frac{1}{\pi(a_i)} \int_{a_i}^{a_{i+1}} b(a)\pi(a)da, \qquad i = 1, 2, \ldots, n \tag{10}$$

Similarly, Equation (4) reduces to

$$\sum_{i=1}^{n} c_i u(a_i+) = 1 \tag{11}$$

where

$$c_i \triangleq \frac{1}{\pi(a_i)} \int_{a_i}^{a_{i+1}} c(a)\pi(a)da, \qquad i = 1, 2, \ldots, n \tag{12}$$

For the yield of the harvest, we note that a total of $u(a_i-) - u(a_i+)$ females per unit time are harvested at age a_i. We thus have

$$\text{Yld} = \sum_{i=1}^{n} y(a_i)[u(a_i-) - u(a_i+)] \tag{13}$$

Next, we see from Equation (8) that

$$u(a_{i+1}-) = u(a_i+)\pi(a_{i+1})/\pi(a_i), \qquad i = 1, 2, \ldots, n-1 \tag{14}$$

Consequently, if we set

$$B_i \triangleq \pi(a_{i+1})/\pi(a_i), \qquad i = 1, 2, \ldots, n-1 \tag{15}$$

then Equations (9) and (14) can be written in matrix form as

$$\begin{bmatrix} u(a_1-) \\ u(a_2-) \\ u(a_3-) \\ \vdots \\ u(a_n-) \end{bmatrix} = \begin{bmatrix} A_1 & A_2 & \cdots & A_{n-1} & A_n \\ B_1 & 0 & \cdots & 0 & 0 \\ 0 & B_2 & \cdots & 0 & 0 \\ \vdots & \vdots & & \vdots & \vdots \\ 0 & 0 & \cdots & B_{n-1} & 0 \end{bmatrix} \begin{bmatrix} u(a_1+) \\ u(a_2+) \\ u(a_3+) \\ \vdots \\ u(a_n+) \end{bmatrix} \tag{16}$$

Let the letter M denote the $n \times n$ matrix on the right-hand side of this equation. Also, let the following n-dimensional column vectors be introduced:

$$u = [u(a_i+)], \quad c = [c_i], \quad y = [y(a_i)]$$

Then Equations (11) and (13) can be written as $c^T u = 1$ and $Yld = y^T(Mu - u)$. The problem of maximizing the yield can now be stated as:

PROBLEM 2 Maximize $y^T(Mu - u)$ subject to

1. $u \geq 0$
2. $Mu \geq u$
3. $c^T u = 1$

The condition $Mu \geq u$ is needed to guarantee that $u(a_i-) \geq u(a_i+)$ at each harvested age. This Linear Programming problem was investigated by Rorres [7]. A similar problem was earlier discussed by Beddington and Taylor [1] and Rorres and Fair [8], though not in a Linear Programming setting. The conclusion drawn by these authors was that the objective function $y^T(Mu - u)$ is maximized when $(Mu - u)_i = 0$ (i.e., $u(a_i-) = u(a_i+)$) for all i with either one or two exceptions. Furthermore, if there are two exceptional values of i, then $u(a_j+) = 0$ for all j from the second exceptional value of i on. In terms of the original impulsive harvesting problem, this means that if impulsive harvesting occurs at ages $a_1, a_2, \ldots, a_n$, the maximum yield can be obtained by harvesting at most two of these ages, with complete harvesting of the population at the older age if it is present.

Having established that the optimal impulsive policy is bimodal, we next attempt to determine the two ages which should be harvested. We do this by computing the yield obtained by impulsive bimodal harvesting at two arbitrary ages a_1 and a_2 ($a_1 \le a_2$) and then finding the values of a_1 and a_2 which maximize the yield. Accordingly, let us set

$Yld(a_1, a_2)$ = yield when the population is harvested partially at age a_1 and completely at age a_2 where $0 \le a_1 \le a_2 \le L$

Here, L denotes the maximum lifetime of any female when no harvesting occurs. With such harvesting, u(a) is given by

$$u(a) = \begin{cases} u(0)\pi(a), & 0 \le a < a_1 \\ u(a_1+)\pi(a)/\pi(a_1), & a_1 < a < a_2 \\ 0, & a_2 < a \le L \end{cases} \tag{17}$$

The condition $u(0) = \int_0^A b(a)u(a)da$ then becomes

$$u(0) = u(0)\int_0^{a_1} b(a)\pi(a)da + \frac{u(a_1+)}{\pi(a_1)}\int_{a_1}^{a_2} b(a)\pi(a)da \tag{18}$$

We next define

$$R(a) \triangleq \int_0^a b(\xi)\pi(\xi)d\xi \tag{19}$$

The function R(a) can be interpreted as the expected number of daughters born to a female in the unharvested population up to age a. Equation (18) can then be rewritten:

$$u(0) = u(0)R(a_1) + u(a_1+)\big(R(a_2) - R(a_1)\big)/\pi(a_1) \tag{20}$$

If we set

θ = fraction of the population harvested at age a_1

it then follows from Equation (20) and the fact that $u(a_1-) = u(0)\pi(a_1)$ that

$$\theta = \frac{u(a_1-) - u(a_1+)}{u(a_1-)} = \frac{R(a_2) - 1}{R(a_2) - R(a_1)} \tag{21}$$

Since we want θ to lie in the interval [0, 1], Equation (21) requires that a_1 and a_2 be such that $0 \leq R(a_1) \leq 1 \leq R(a_2)$. Using the fact that R(a) is monotonically increasing, this condition can also be written as $0 \leq a_1 \leq \hat{a} \leq a_2$ where $\hat{a}$ is the age at which $R(\hat{a}) = 1$. Thus $\hat{a}$ can be interpreted as the *replacement age*: that age at which a female first replaces herself by having given birth to an average of one daughter.

From the constraint condition $1 = \int_0^A c(a)u(a)da$ we obtain, similar to Equation (20):

$$1 = u(0)C(a_1) + u(a_1+)\big(C(a_2) - C(a_1)\big)/\pi(a_1) \tag{22}$$

where we define

$$C(a) \triangleq \int_0^a c(\xi)\pi(\xi)d\xi \tag{23}$$

It can be seen that C(a) is the expected cost of raising one female to age a. For the yield $Yld(a_1, a_2)$ we have

$$Yld(a_1, a_2) = y(a_1)[u(a_1-) - u(a_1+)] + y(a_2)u(a_2-) \tag{24}$$

Combining Equations (21), (22), and (24) then gives

$$Yld(a_1, a_2) = \frac{Y(a_2) - Y(a_1)}{X(a_2) - X(a_1)} \tag{25}$$

where

$$X(a) \triangleq C(a)/[1 - R(a)]$$

$$Y(a) \triangleq y(a)\pi(a)/[1 - R(a)]$$

Thus once a_1 and a_2 are specified, the yield $Yld(a_1, a_2)$ is uniquely determined by Equation (25) through the specified functions $y(a)$, $\mu(a)$, $b(a)$, and $c(a)$. The problem of finding the optimal bimodal impulsive harvesting policy reduces to finding the maximum value of $Yld(a_1, a_2)$ when a_1 and a_2 vary over the rectangular region $\{0 \leq a_1 \leq \hat{a} \leq a_2 \leq L\}$.

A geometric interpretation of finding the maximum of $Yld(a_1, a_2)$ is given in Figure 11.1. There we have plotted the points $(X(a), Y(a))$ as a varies from 0 to L. The resulting parametric curve lies in the first quadrant for $a \in [0, \hat{a})$ and in the third quadrant for $a \in (\hat{a}, L]$. Specifying the two harvesting ages a_1 and a_2 satisfying $0 \leq a_1 \leq \hat{a} \leq a_2 \leq L$ thus corresponds to selecting one point on the curve in the first quadrant and another point on the curve in the third quadrant. From Equation (25), the resulting yield is numerically equal to the slope of the line passing through those two points. The maximum yield then corresponds to the line of the indicated form with maximum slope.

As indicated in Figure 11.1, the line of maximum slope is usually tangent to the curve $(X(a), Y(a))$ at the points which determine the optimal harvested ages. In Figure 11.2, four degenerate cases are illustrated, corresponding to "endpoint" solutions, where such tangency need not necessarily take place.

RESTRICTED HARVESTING

In this section we make a few remarks about the case when the condition $h(a) \geq 0$ in Problem 1 is replaced by $0 \leq h(a) \leq H$ where H is a fixed constant. With an upper bound on the harvesting rate, impulsive harvesting is no longer possible. To treat this problem, we introduce the Hamiltonian $\tilde{H}(a)$ and the costate variables $\eta_1(a)$, $\eta_2(a)$, and $\eta_3(a)$, as determined by the following equations:

$$\tilde{H}(a) = y(a)h(a) + \eta_1(a)[-\mu(a)u(a) - h(a)] + \eta_2(a)b(a)u(a) + \eta_3(a)c(a)u(a)$$

$$\eta_1'(a) = -\partial\tilde{H}/\partial u = \mu(a)\eta_1(a) - b(a)\eta_2(a) - c(a)\eta_3(a)$$

$$\eta_2'(a) = -\partial\tilde{H}/\partial v = 0 \qquad (26)$$

$$\eta_3'(a) = -\partial\tilde{H}/\partial w = 0$$

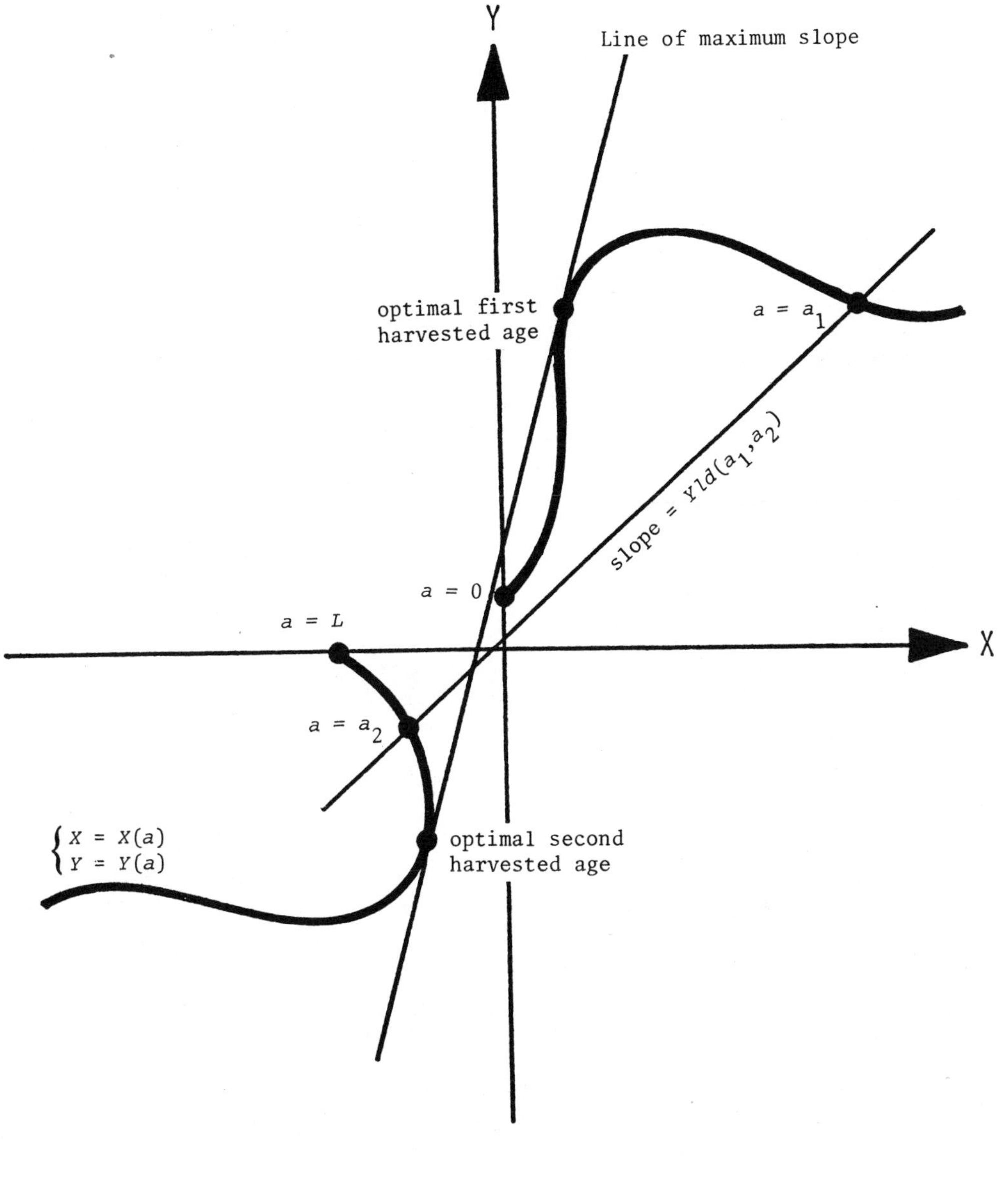
Y
Line of maximum slope
optimal first
harvested age
a = a1
slope = Yld(a1,a2)
a = 0
a = L
X
a = a2
X = X(a)
Y = Y(a)
optimal second
harvested age

Fig. 11.1.

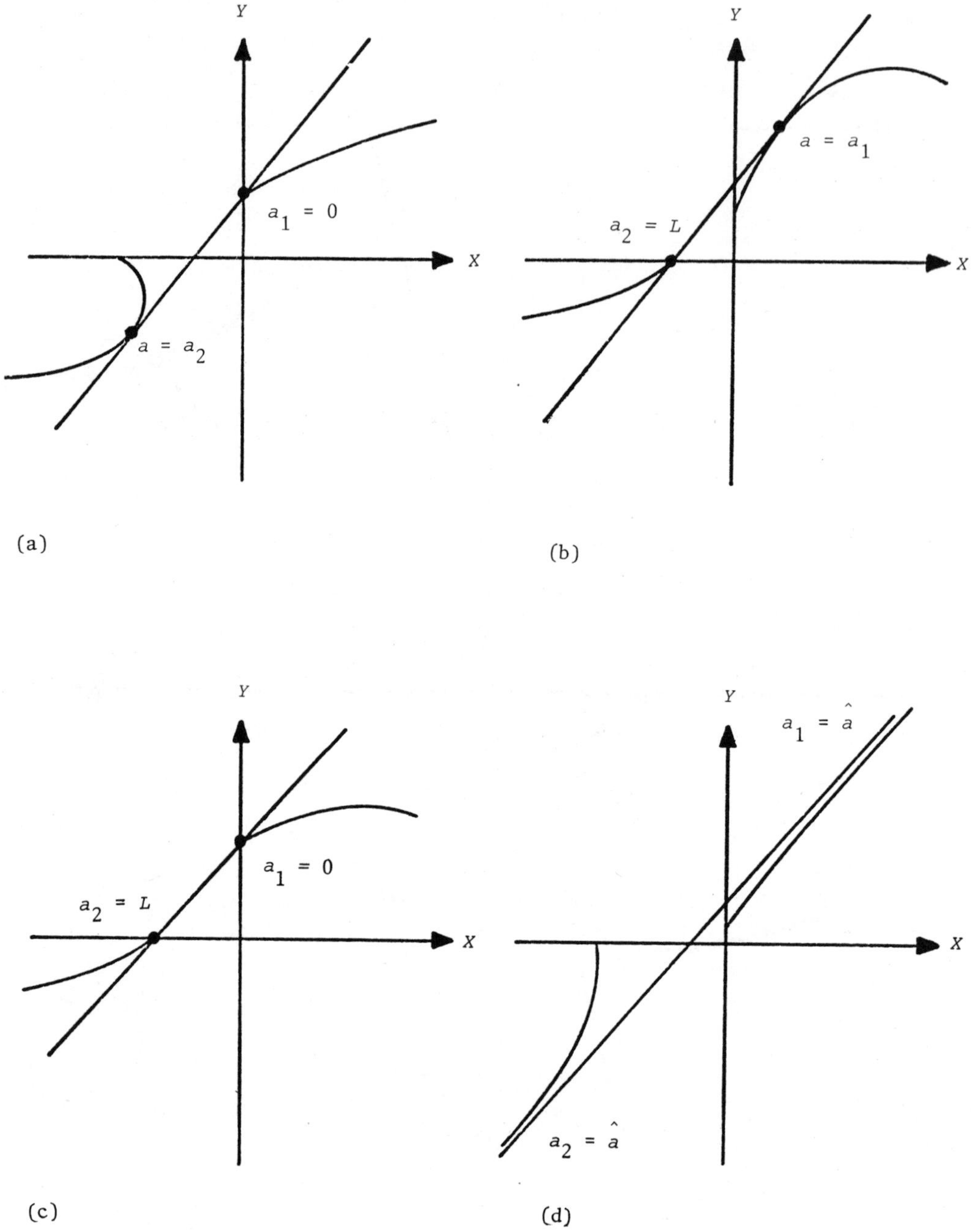

Fig. 11.2.

The transversality conditions at the endpoints reduce to $\eta_1(0) = \eta_2(0)$ and $\tilde{H}(A) = 0$. The maximum principle requires that $\tilde{H}(a)$ be maximized at each $a \in [0, A]$, from which it follows that

$$h(a) = \begin{cases} 0 & \text{if } y(a) < \eta_1(a) \\ H & \text{if } y(a) > \eta_1(a) \end{cases} \tag{27}$$

From the second and third costate equations (26) it follows that $\eta_2(a)$ and $\eta_3(a)$ are constants. Setting $\eta_2(a) \equiv K_0$ and $\eta_3(a) \equiv -K$, substituting these values into the first costate equation (26), and solving the resulting first order linear ordinary differential equation for $\eta_1(a)$ subject to the initial condition $\eta_1(0) = \eta_2(0) = K_0$, we obtain

$$\eta_1(a)\pi(a) = K_0[1 - R(a)] + KC(a) \tag{28}$$

where $\pi(a)$, $R(a)$, and $C(a)$ were previously defined by Equations (7), (19) and (23).

From the fact that $u(A) = 0$, the transversality condition $\tilde{H}(A) = 0$ reduces to

$$h(A)[y(A) - \eta_1(A)] = 0 \tag{29}$$

Now if $h(A) = 0$, i.e., no harvesting occurs at the oldest age A attained by any female in the harvested population, then it must be that A is in fact L. Since $\pi(L) = 0$, Equation (28) becomes at $a = L$

$$0 = K_0[1 - R(L)] + KC(L) \tag{30}$$

On the other hand, if $h(A) \neq 0$, then Equation (29) requires that $y(A) = \eta_1(A)$. Substituting this into Equation (28) at $a = A$ yields

$$y(A)\pi(A) = K_0[1 - R(A)] + KC(A) \tag{31}$$

It is seen that Equation (30) is just the special case of Equation (31) when $A = L$.

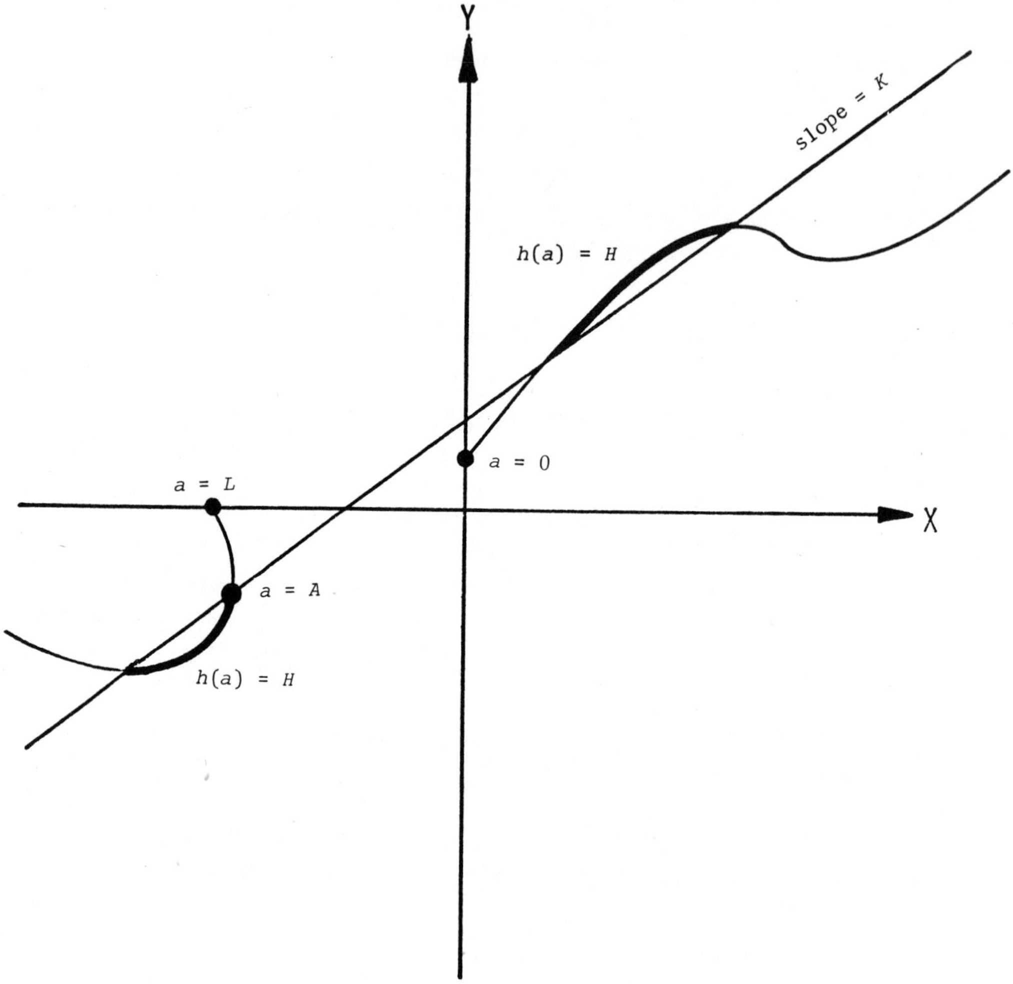

Fig. 11.3

Eliminating K_0 between Equations (28) and (31) and solving for the switching function $\eta_1(a)$ then gives

$$\eta_1(a) = \frac{y(A)\pi(A)[1 - R(a)] + K\{C(a)[1 - R(A)] - C(A)[1 - R(a)]\}}{\pi(a)[1 - R(A)]} \qquad (32)$$

Substituting this into the inequality $y(a) > \eta_1(a)$ and rearranging the terms a bit then results in the following characterization of the optimal harvesting policy according to the switching criterion in Equation (27):

The optimal harvesting policy is such that at each age either $h(a) = 0$ *or* $h(a) = H$. *Maximum harvesting* $h(a) = H$ *occurs at those ages for which*

$$\frac{Y(a) - Y(A)}{X(a) - X(A)} > K$$

where K *is a constant and* A *is the maximum age attained by any female in the harvested population.*

Figure 11.3 provides the following geometric interpretation of this result: In the X-Y plane containing the parametric curve $(X(a), Y(a))$, $0 \le a \le L$, there is a line of slope K passing through the point $(X(A), Y(A))$ such that harvesting takes place maximally only for those ages corresponding to the portions of the curve cut off by this line. In the diagram these portions of the curve for which $h(a) = H$ are indicated by a heavy line.

It can be seen that as $H \to \infty$, the line of slope K approaches the line of maximum slope determined by the optimal bimodal impulsive harvesting problem. Consequently, for H sufficiently large, the optimal policy consists of harvesting two age intervals which contain the two optimal harvested ages a_1 and a_2 of the previous section.

REFERENCES

1. J. R. Beddington and D. B. Taylor, "Optimal age specific harvesting of a population," *Biometrics*, 29(1973), 801-809.

2. F. Brauer, "Constant rate harvesting of populations governed by Volterra integral equations," *J. Math. Anal. Appl.* 56(1976), 18-27.

3. C. W. Clark, *Mathematical Bioeconomics*. John Wiley and Sons, New York (1976).

4. W. G. Doubleday, "Harvesting in matrix population models," *Biometrics* 31(1975), 189-200.

5. L. P. Lefkovitch, "The study of population growth in organisms grouped by stages," *Biometrics* 21(1965), 1-18.

6. L. P. Lefkovitch, "A theoretical evaluation of population growth after removing individuals from some age groups," *Bull. ent. Res.* 57(1966), 437-445.

7. C. Rorres, "Optimal sustainable yield of a renewable resource," *Biometrics* 32(1976), 945-948.

8. C. Rorres and W. Fair, "Optimal harvesting policy for an age-specific population," *Math. Biosci.* 24(1975), 31-47.

9. D. A. Sánchez, "Linear age-dependent population growth with harvesting," *Bull. Math. Biol.* 40(1978), 377-385.

MODELS INVOLVING DIFFERENTIAL AND INTEGRAL EQUATIONS APPROPRIATE FOR DESCRIBING A TEMPERATURE DEPENDENT PREDATOR-PREY MITE ECOSYSTEM ON APPLES

David J. Wollkind

Department of Pure and Applied Mathematics
Washington State University
Pullman, Washington

Alan Hastings

Department of Pure and Applied Mathematics
Washington State University
Pullman, Washington

Jesse A. Logan

The Natural Resource Ecology Laboratory
and the Department of Zoology and Entomology
Colorado State University
Fort Collins, Colorado

INTRODUCTION

The McDaniel spider mite, *Tetranychus mcdanieli* McGregor, has occurred on apple tree foliage in the Wenatchee area of Washington State ever since such orchards were first introduced in the interior valleys of that state. Initially this phytophagous mite was not considered a major pest; light summer oils (added to lead arsenate codling moth sprays) were usually enough to provide adequate control. Even when high populations did occur, they were typically late in the growing season and caused little damage [13]. Pest

Current affiliation:
Dr. Hastings is presently at the Department of Mathematics, University of California, Davis, California.

status of this species was amplified by the first widespread use of DDT for codling moth control during the 1947 growing season.

The impact of DDT on tetranychid mite populations was twofold. First, because it provided an effective means for controlling codling moth, damage caused by these spider mites now became more noticeable to growers. Secondly, DDT directly affected mite populations by drastically reducing the level of the predacious phytoseiid mite *Metaseiulus occidentalis* Nesbitt which feeds upon McDaniel mites. An era of intensive *chemical control* followed the introduction of DDT. During this period, success of such control for McDaniel spider mites was uncertain. Typically, resistance developed to even those chemicals which initially provided adequate control [5].

By the early 1960s, McDaniel spider mites constituted a significant problem with high populations occurring early in the summer and economic damage common. This general situation continued to deteriorate until a successful *integrated mite control* program was introduced by Hoyt [5] in 1965.

The framework for integrated mite control was established when Hoyt observed a case in which a predatory mite survived a codling moth chemical cover spray program. Subsequent studies resulted in a comprehensive control program integrating chemical control of insect pests with the *biological control* of tetranychid mites on apples. Biological control of *T. mcdanieli* by *M. occidentalis* is based on maintaining a ratio of phytophagous to predacious mites favorable to achieving adequate control. If the time lag between prey build-up and eventual suppression by the predator is great enough to allow economically important densitites to develop or if early season over-exploitation results in predator starvation and subsequent build-up of the prey to economic levels, then such control will fail [5]. The critical importance of temperature on the interaction between *T. mcdanieli* and *M. occidentalis* has long been recognized and a good deal of experimental effort has been expended determining the temperature dependence of the parameters involved [11, 12]. It is the development of models relevant to this temperature-dependent two-species interaction on apple tree foliage with which this paper is concerned (see Figure 12.1). In the next section we present a simulation model involving differential equations while in the third section a linear stability analysis is performed on the critical points of a modified system which contains an integral equation as well.

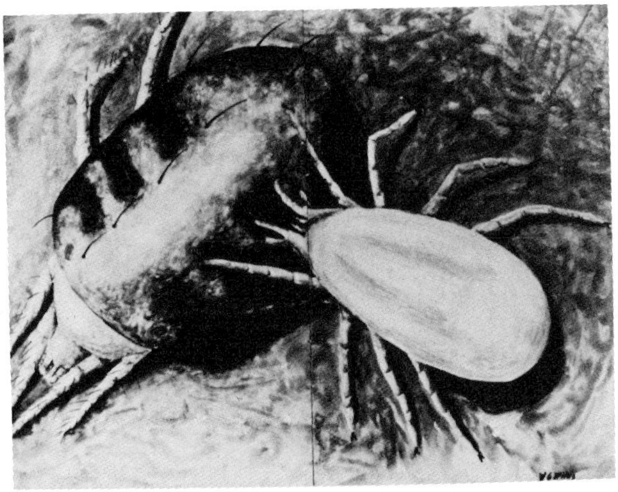

Fig. 12.1. An artist's (W. G. Wiles) impression of the attack by the predacious mite *Metaseiulus occidentalis* on the phytophagous spider mite pest *Tetranychus mcdanieli* occurring upon an apple leaf (courtesy of the Fred L. Overly Laboratory, Tree Fruit Research Center, Washington State University, Wenatchee, Washington).

A DIFFERENTIAL EQUATIONS SIMULATION MODEL

Consider a two-component ecosystem which consists of a single carnivorous mite species feeding exclusively on a single herbivorous one. Since the time period of interest is from mid to late summer when the mites are relatively numerous while the field sampling technique to be employed for validation purposes assumes uniform mite distribution, we first model this interaction mathematically by introducing two differentiable functions of a continuous time variable (denoted by t): namely

$$H(t) \equiv \text{population density of the herbivore } (\textit{T. mcdanieli})$$

and

$$C(t) \equiv \text{population density of the carnivore } (\textit{M. occidentalis})$$

both measured in number of adult females per leaf, which satisfy the following basic equations governing the growth of these populations:

$$\frac{dH}{dt} = r_1(T)H - \phi(H, C; T)C \tag{1a}$$

and

$$\frac{dC}{dt} = -g(\phi)C + r_2(T)h(\phi)C \quad \text{for} \quad t > 0 \tag{1b}$$

with initial conditions at t = 0 of

$$H(0) = H_0 \quad \text{and} \quad C(0) = C_0 \tag{1c}$$

where ϕ describes the predator *functional* response to the prey--i.e., the number of prey killed per unit time by a single average predator; $h(\phi)$, the predator *numerical* response or fecundity; and $g(\phi)$, the predator *starvation* response; while $r_1(T)$ and $r_2(T)$ are the *intrinsic growth rates* of the prey and predator population with unlimited food supply as functions of temperature, T, which is itself a function of time--i.e., $T = T(t)$; and H_0 and C_0 are initial population densities.

From the outset the model has been restricted to the *female component* of both species. This restriction implicitly assumes that enough males are present in the population to assure mating and also that males are of minor importance compared to females, with respect to economic impact on the apple crop or as biological control agents. Also implicit in the formulation of system (1) is the assumption that the time scale of interest is long when compared with the scale over which time lags occur but short in comparison with the scale at which genetic mutations might become significant. In addition, mite metamorphosis from egg through immature stages to adult while not considered explicitly is included implicitly during the determination of the parameters contained in the intrinsic growth rates which follows.

Unlike many models for insect population dynamics which introduce temperature stochastically [10], these temperature effects have been approached deterministically in the model of (1). In particular it is necessary to find an analytic expression as a function of temperature for $r_1(T)$ and a similar one for $r_2(T)$. The effect of temperature on each of these rates is divided into two phases. During Phase I that growth rate can be characterized as increasing monotonically to a maximum with temperature. During Phase II (above this optimum temperature) it undergoes a rapid (often precipitous) decline to zero at the lethal maximum temperature.

Motivated both by the occurrence of this *boundary layer* type behavior in the high temperature range and by the form [7] of the solution $y(x; \varepsilon)$ for the singularly perturbed problem $\varepsilon y'' + y' + by = 0$, $y(0; \varepsilon) = 0$, $y(1; \varepsilon) = 1$, one can deduce an analytic expression for the intrinsic growth rate of either mite species as a function of temperature. Denoting this rate per day by $r(T)$ where T is measured in °C above a given base temperature, it can be shown to be of the form [7]

$$r(T) = \psi\{\exp(\rho T) - \exp[\rho T_M - (T_M - T)/\Delta T]\} \quad \text{for} \quad T \in [0, T_M] \tag{2}$$

where the biologically meaningful parameters ψ, ρ, T_M, and ΔT are to be identified by a least squares fit to the relevant *generated* data. In particular, to determine the effect of temperature on the dynamics of *T. mcdanieli*, it was necessary to incorporate the temperature-rate equation of (2) into a synoptic model describing a *complete* life history. This was accomplished by formulating a Forrester [2]-type discrete time dynamical systems model using Equation (2) with proper parameter identifications to update each life stage as well as to compute oviposition [6]. The maximum population growth rate possible for adult McDaniel mites at a given set of temperatures, $r_m(T_k)$, can be estimated from this discrete time model by computing the adult population growth rate $r(T_k)$ at each time step and allowing enough time for convergence of $r(T_k)$ to $r_m(T_k)$. A least squares fit of (2) to these generated data points finally results in the following determination of $r_1(T)$:

$$r_1(T) = .048\{\exp(.103T) - \exp[2.89 - (28.04 - T)/2.71]\} \tag{3}$$

for T = °C - 10° C

Although we could have determined $r_1(T)$ directly from observed data for the adult stage, this approach besides being difficult to accomplish, suffers from an additional basic deficiency. Namely, it does not take the actual life history of the McDaniel mite into account.

Proceeding in a similar manner to the one outlined above for the derivation of $r_1(T)$, it is possible to obtain an explicit representation for $r_2(T)$ of the same generic form as (3) from an analogous parameter identification of the relevant generated data points for *M. occidentalis* [6]. Both these functions are represented in Figure 12.2 where the solid line is $r_1(T)$ while the broken one is $r_2(T)$.

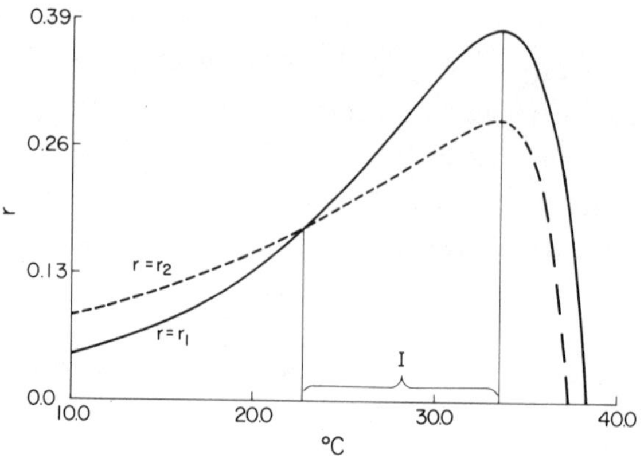

Fig. 12.2. A comparison of the simulated r_m values as a function of temperature for the two mite species. The solid line is $r_1(T)$ for *T. mcdanieli* while the broken line is $r_2(T)$ for *M. occidentalis*. The interval I is being designated for future reference in the third section.

We now develop appropriate expressions for the related functions ϕ, h, and g contained in (1). Many authors have considered the functional response, ϕ, to be independent of the predator and have taken it to be a function of prey density alone [10]. In the simplest model of this sort, the Lotka-Volterra model [8]

$$\phi(H, C) = aH \tag{4}$$

which implies that the predator never becomes satiated since its rate of consumption increases linearly as new individual prey are introduced. Such an assumption is invalid for most natural systems. Holling [4] improved on this by using an expression of the form

$$\phi(H, C) = \frac{aH}{H + b_1} \tag{5}$$

which fit experimental results for invertebrate predators well. Although we shall employ the Holling functional response of (5) in our model to be presented in the next section, we now introduce a ϕ which is a function of the ratio of the population densities rather than of just the prey density. That is, upon suppression of the temperature dependence contained in (1a), we take

$$\phi(H, C) = af(\nu) \quad \text{where} \quad \nu = H/C \tag{6}$$

such that $C \neq 0$ while $H = 0$ corresponds to $\nu = 0$. Using a Holling-type approach, $f(\nu)$ is defined by

$$f(\nu) = \frac{\nu}{\nu + b_1} \quad \text{where} \quad b_1 = \tilde{a} = \frac{a^*}{1 + r_1^*} \tag{7}$$

for a^* and r_1^* having the numerical value of a and r_1 respectively but possessing their corresponding dimensionality multiplied by one day. Note such an f satisfies the conditions (see Figure 12.3)

$$f(0) = 0, \quad \lim_{\nu \to \infty} f(\nu) = 1, \quad \text{and} \quad f(\nu) \leq \nu/\tilde{a} \tag{8a, b, c}$$

That condition (8c) is necessary, is a direct consequence of the discrete time condition [9]

$$(\Delta t)a\ f(\nu)C \leq H + (\Delta t)\ r_1 H$$

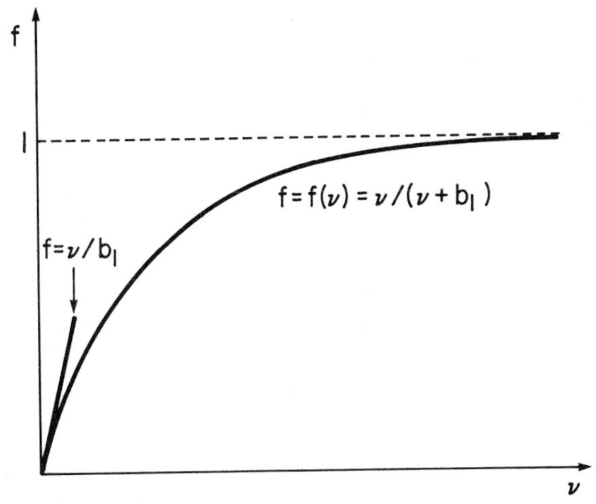

Fig. 12.3. A graph of the functional response curve of (7) with the tangent line to it at the origin denoted explicitly.

with Δt taken equal to one day, which guarantees that the predator in any time interval (in this case, one day) can kill no more prey than are in the population. For discrete time models, condition (8c) ensures that negative populations do not arise during the simulation and has been adopted in this continuous time case for the sake of uniformity since there is some merit in having a general system which can be applied equally well to both discrete and continuous time situations. Observe that, while the Holling functional response of (5) implies $\phi \to a$ (maximum rate of consumption) as $H \to \infty$ independent of C, this only occurs for our model of (7) provided $H/C \to \infty$. In other words, the rate of consumption per predator becomes maximal not when the number of prey grows without bound but only when that quantity per predator does.

Finally, it is necessary to deduce explicit representations for the functions h and g included in (1). Since the rate of predator fecundity (numerical response) and rate of predator starvation (starvation response) are dependent upon nutrition or the lack of it, h and g have been considered functions of ϕ. In particular, it is expected that

$$r_2h(\phi) \to r_2 \quad \text{and} \quad g(\phi) \to 0 \quad \text{as} \quad \phi \to a \tag{9a}$$

while

$$r_2h(\phi) \to 0 \quad \text{and} \quad g(\phi) \to c \quad \text{as} \quad \phi \to 0 \tag{9b}$$

where c is the maximum predator starvation rate. Hence h and g of the form

$$h(\phi) = n(f) = \frac{f}{1 + b_0(1 - f)} \tag{10a}$$

and

$$g(\phi) = s(f) = c\left[1 - \frac{f}{1 + d_0(1 - f)}\right] \tag{10b}$$

are introduced where b_0 and d_0 are "shaping constants" assumed positive.

We now incorporate the expressions developed above for the intrinsic growth rates and predator responses into our basic system (1). In addition, a and c are considered fixed and estimated from environmental chamber studies while the "shaping constants" have been given appropriate values (see below for a further discussion of these parameters). Then we obtain numerical solutions for this system by merging the relevant temperature-time data [6] with a continuous numerical differential equation solver. Temperature adjustments are made at hourly intervals and the solver restarted at the beginning of each interval. A phase plane plot comparing the results of this model to 1967 field data is given in Figure 12.4 which represents a thirty-day time interval with each numbered point denoting two and one-half days duration. Although differences in response between these results and the observed data do exist, it should be noted that *qualitatively* the responses are quite similar. That is, the prey population builds to a maximum and then "crashes" taking the predator to extinction with it.

The parameters a and c which have been taken as fixed are actually functions of temperature. In work completed subsequently to that described in this section, we have included a temperature dependent relationship for a derived from relevant data in a manner similar to that employed earlier to determine r_1 and r_2 as functions of temperature. Such analysis did not significantly alter the qualitative behavior of the system and even the

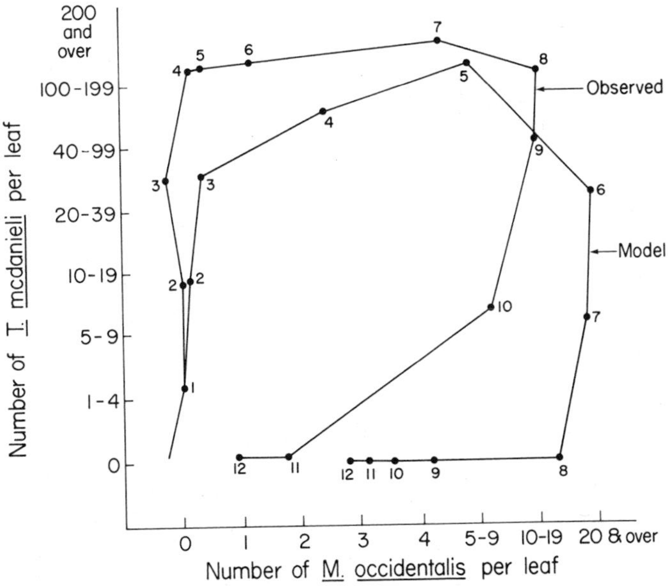

Fig. 12.4. A phase plane plot comparing the results of the simulation model with observed field data.

quantitative effect caused by this inclusion was slight. Hence we can state with some confidence that the temperature dependent relationships of importance to the population dynamics of the system are the ones shown in Figure 12.2. As far as c goes, this parameter in effect determines the duration of the residual predator population after overexploitation has occurred. For our specific application of this section, the residual persistence of the predator in the absence of the prey is *not* of overriding importance (see the third section). What is of importance is the *buildup* of the populations at densities which can be sampled in the field. Hence all we needed for the purposes of our model was a response that starved the predator in an empirically reasonable fashion and a constant c satisfied this condition. The presence of both "shaping parameters" b_0 and d_0 came about by assuming a Michaelis-Menten type hyperbolic relationship between nutritional level and proportion of maximum population growth achieved. While this is an interesting, and somewhat logical conjecture, it was beyond the intent of this study to determine exactly the nature of that relationship. Again, parameters which resulted in empirically reasonable model behavior were selected by trial and error. For our analysis b_0 and d_0 were assigned the common value of 0.5. In this range the model seemed relatively insensitive to changes in b_0 or d_0. We note that all of these parameters are of biological interest but explicit determination of any one of them could comprise a major research effort. The importance of this model, as we see it, is to provide a conceptual framework and a mechanism for testing the effect of including temperature dependence in the intrinsic growth rates.

The purpose of the model is to quantify the phenomena in question sufficiently to provide growers with adequate information to enable them to plan control strategies. In particular, for this problem the apple grower is faced with a given initial population of the pest species, the McDaniel spider mite. He wishes to know if the existing density of predacious mites is at the right level to provide adequate biological control of the type discussed in the Introduction. This, in turn, depends on both the maximum prey density achieved and the time it takes the whole ecosystem to crash. It may be possible to achieve such control by adding predacious mites initially, if without this addition the prey maximum would be too large, or by conserving phytophagous mite populations if the system crashes too early. Perhaps the most important short-term question such a model can answer, therefore, is, given the number of predator and prey present, how long and at what level will *T. mcdanieli* populations be maintained. In other words, will chemical control become necessary?

The model as presented in this section has provided useful information about the dynamic structure of the field populations and also served as a first step toward the development of a comprehensive model describing orchard mite populations in Washington State. Once this has been completely achieved, it will then be possible to formulate a simulation model to predict the economic damage inflicted on the orchards by phytophagous mites (see Figure 12.5). For our simplified model (see the third section) this might be attempted by appending to the original system of (1) an additional differential equation governing apple size S(t), of the form

$$\frac{dS}{dt} = k(T) - p(H)S - q(H) \text{ for } t > 0,\ S(0) = S_0 \tag{11}$$

where p, q, and k are positive increasing functions of their respective arguments.

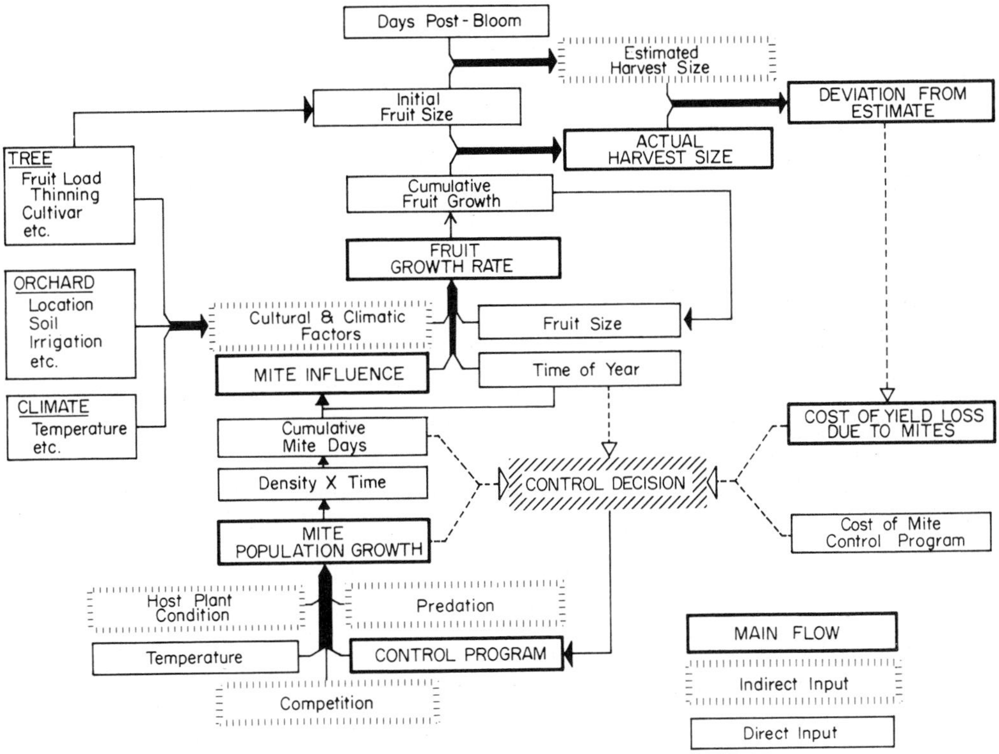

Fig. 12.5. A schematic flow chart, due to L. K. Tanigoshi, representative of a complete simulation model for determining the economic influence of mites on apple orchards.

A LINEAR STABILITY ANALYSIS RELEVANT TO A MODEL CONTAINING AN INTEGRAL EQUATION

Obviously the model presented in the previous section could be improved upon by including a number of somewhat complex factors that were neglected during its formulation. Spatially dependent, diffusion-like migration effects could have been considered, although it should be noted that the field sampling procedure used to obtain data for comparative evaluation of model behavior also required the assumption that the species were homogeneously distributed in space. Further the model might have been modified to account for the fact that if alternate prey species, primarily the apple rust mite, *Aculus Schlectendali* Nalepa, are available in early summer, biological control of *T. mcdanieli* by *M. occidentalis* is virtually assured. These alterations would require that we introduce additional independent and dependent variables in our original model of the second section. In this section, however, we shall modify that model without recourse to such mathematical complications as a first step toward improving its quantitative accuracy and then perform a linear stability analysis of the critical points of the resulting system.

We begin by replacing (1a) with the equation

$$\frac{dH}{dt} = r_1 H\left[1 - \frac{H}{K} - \frac{aC}{r_1(H + b_1)}\right], \quad t > 0 \qquad (12a)$$

where we have introduced a finite spider mite carrying capacity, K, which was assumed to be infinite previously, and adopted the Holling functional response of (5). Next we replace the starvation response of our original model by a more realistic mechanism which accounts for the fact that in the absence of its prey the carnivore simply dies off. This can be accomplished by adopting the integral equation

$$C(t) = \int_{t-D}^{t} \left[\frac{r_2 HC}{H + b_2}\right](s)ds, \quad t > 0 \qquad (12b)$$

where D(T) is the ovipositional period of the carnivore as a function of temperature and $b_2 = b_1(1 + b_0)$, instead of (1b). Equation (12b) is developed as follows: Consider the age structure Von Foerster [1]-type continuity equation for $\tilde{C}(A, t) \equiv$ population density of carnivores of age A at time t, given by

$$\frac{\partial \tilde{C}}{\partial t} + \frac{\partial \tilde{C}}{\partial A} = 0, \qquad 0 < A < D, \quad t > 0 \tag{13a}$$

$$\tilde{C}(0, t) = [r_2 h(\phi) C](t) \equiv \text{birth rate}, \qquad t > 0 \tag{13b}$$

where ϕ is the functional response of (5) and h, the numerical response of (10a). The population density C(t) of the carnivore at any time t is related to $\tilde{C}(A, t)$ through the equation

$$C(t) = \int_0^D \tilde{C}(A, t)\,dA \tag{14}$$

Thus solving (13) by the method of characteristics to yield

$$\tilde{C}(A, t) = [r_2 h(\phi) C](t - A) \tag{15}$$

substituting (15) into (14), and making the change of variables $s = t - A$, we obtain the integral equation of (12b). In addition it now becomes necessary to stipulate H and C for the initial time interval $t \in [-D, 0]$ rather than just at the initial time $t = 0$ as was the case previously in (1c). That is we prescribe

$$H = H_0(t; T) \quad \text{and} \quad C = C_0(t; T) \quad \text{for} \quad t \in [-D, 0] \tag{16}$$

where $H_0(t; T)$ and $C_0(t; T)$ are initial population density functions.

Note that such an approach models starvation by actually eliminating it as a fictitious artifice of Lotka-Volterra type models. In the absence of the herbivore for this model the carnivore just stops reproducing since (12b) contains the numerical response. Then after a time period of D(T) all the carnivores die of natural mortality rather than starve to death. Since this is a valid behavioral tendency when dealing with adult female anthropod predators, our revised model seems reasonable as a first attempt.

We now turn to the details of the stability analysis mentioned above. There exist three critical points of the system of (12). Denoting such steady state solutions of (12) by

$$H(t) = H^* \quad \text{and} \quad C(t) = C^* \tag{17}$$

we find that two

$$H^* = C^* = 0 \quad \text{and} \quad H^* = K, \quad C^* = 0 \tag{18ab}$$

contain a zero carnivore component, which was the purpose of adopting the functional response of (5) instead of that of (7), while one

$$H^* = \frac{b_2}{r_2 D - 1} = H_0 > 0, \quad C^* = \left(\frac{r_1}{a}\right)(H_0 + b_1)\left(1 - \frac{H_0}{K}\right) = C_0 > 0 \tag{18c}$$

represents a community equilibrium point. In this section, for the sake of simplicity, we are considering the temperature T to be a parameter rather than a function of time; hence the r_1, r_2, and D contained in (18c) should be interpreted in that sense. Further we observe that to insure H_0, $C_0 > 0$ and $H_0 < K$, these parameters must satisfy the relation

$$D\, r_2 > 1 + \frac{b_2}{K} \tag{19}$$

In order to determine conditions under which there can exist a stable community equilibrium point for this model we perform a linear stability analysis of the critical point of (18c). First we introduce the following nondimensional variables and parameters:

$$\tau = \frac{t}{D}, \quad h(\tau) = \frac{H(t)}{H_0}, \quad c(\tau) = \frac{C(t)}{C_0}, \quad \sigma_1 = D\, r_1, \quad \sigma_2 = D\, r_2 \tag{20}$$

$$\alpha = \frac{H_0}{K}, \quad \beta = \frac{\sigma_2 - 1}{1 + b_0} = \frac{b_1}{H_0}, \quad \text{and} \quad \gamma = \alpha\beta = \frac{b_1}{K}$$

which transform the system of (12) into

$$\frac{dh}{d\tau} = \sigma_1 h[1 - \alpha h - (1 - \alpha)(1 + \beta)c/(h + \beta)] = F(h, c) \tag{21a}$$

and

$$c(\tau) = \int_{\tau-1}^{\tau} \left[\frac{\sigma_2 ch}{h + \sigma_2 - 1}\right](s)ds = \int_{\tau-1}^{\tau} [G(h, c)](s)ds \tag{21b}$$

where condition (19) is equivalent to

$$\alpha < 1 \quad \text{or} \quad \beta > \gamma \tag{22}$$

Noting that the community equilibrium point of (18c) corresponds to

$$h(\tau) = c(\tau) = 1 \tag{23}$$

which is, of course, a critical point for our new system of (21), we now consider a solution of that system of the form

$$h(\tau) = 1 + \varepsilon x(\tau) + 0(\varepsilon^2), \quad c(\tau) = 1 + \varepsilon y(\tau) + 0(\varepsilon^2) \tag{24}$$

where $|\varepsilon| << 1$.

Substituting (24) into (21); expanding F and G in Taylor series about $h = c = 1$; making use of the fact that $F(1, 1) = 0$ while $G(1, 1) = 1$; neglecting terms of $0(\varepsilon^2)$ and then cancelling the resulting common factor of ε; we obtain the following set of perturbation equations:

$$\frac{dx}{d\tau} = (F_1)_0 x + (F_2)_0 y$$

$$y(\tau) = (G_1)_0 \int_{\tau-1}^{\tau} x(s)ds + (G_2)_0 \int_{\tau-1}^{\tau} y(s)ds \tag{25}$$

where partial derivatives with respect to h and c have been designated by 1 and 2 respectively while the 0 subscript denotes quantities evaluated at $h = c = 1$. That is, in this instance,

$$(F_1)_0 = \frac{\sigma_1(1 - 2\alpha - \gamma)}{1 + \beta}, \quad (F_2)_0 = \sigma_1(\alpha - 1)$$

$$(G_1)_0 = \frac{\sigma_2 - 1}{\sigma_2}, \quad (G_2)_0 = 1 \tag{26}$$

We look for solutions of (25) of the form

$$[x, y](\tau) = [A, B]e^{\lambda\tau} \tag{27}$$

where λ is the growth rate of the perturbation quantities while A and B are constants which could in principle be determined from given initial values for these perturbations. Substituting (27) into (25) we obtain two linear homogeneous equations in the two constants A and B. This is an *eigenvalue* problem for λ with associated eigenvector [A, B]. In order for there to be a nontrivial solution the determinant of these coefficients must vanish. That is

$$\begin{vmatrix} (F_1)_0 - \lambda & (F_2)_0 \\ (G_1)_0\kappa(\lambda) & (G_2)_0\kappa(\lambda) - 1 \end{vmatrix} = 0$$

or

$$\lambda - (G_2)_0\lambda\kappa(\lambda) + [(F_1)_0(G_2)_0 - (F_2)_0(G_1)_0]\kappa(\lambda) - (F_1)_0 = 0 \tag{28a}$$

where

$$\kappa(\lambda) = \begin{cases} (1 - e^{-\lambda})/\lambda & \lambda \neq 0 \\ 1 & \lambda = 0 \end{cases} \tag{28b}$$

is a continuous function of λ since $\lim_{\lambda\to 0} \kappa(\lambda) = 1$. Upon the substitution of (26) into (28), that equation becomes

$$\lambda - \lambda\kappa(\lambda) + c\kappa(\lambda) + d = 0 \tag{29a}$$

where

$$d = \frac{\sigma_1(\gamma + 2\alpha - 1)}{1 + \beta} \quad \text{and} \quad c = -d + \sigma_1(1 - \alpha)(\sigma_2 - 1)/\sigma_2 \tag{29b}$$

Since

$$\lim_{\tau\to\infty} |e^{\lambda\tau}| = \lim_{\tau\to\infty} e^{(\mathrm{Re}\lambda)\tau} = \begin{cases} 0 & \mathrm{Re}\lambda < 0 \\ 1 & \mathrm{Re}\lambda = 0 \\ \infty & \mathrm{Re}\lambda > 0 \end{cases}$$

we say the solution of (21) given by the critical point of (23) is stable, neutrally stable, or unstable to the type of perturbations of (24) according to whether Reλ is less than, equal to, or greater than zero. The stability of secular equations of the form of (29) has been analyzed by Hastings [3]. The results of that analysis are summarized in Figure 12.6. In this figure the curve $c = -d$ which corresponds to $\lambda = 0$ in (29) separates the region where community equilibrium can exist in the sense that (22) is satisfied from the unattainable region, $c + d < 0$, where it can't. In the attainable

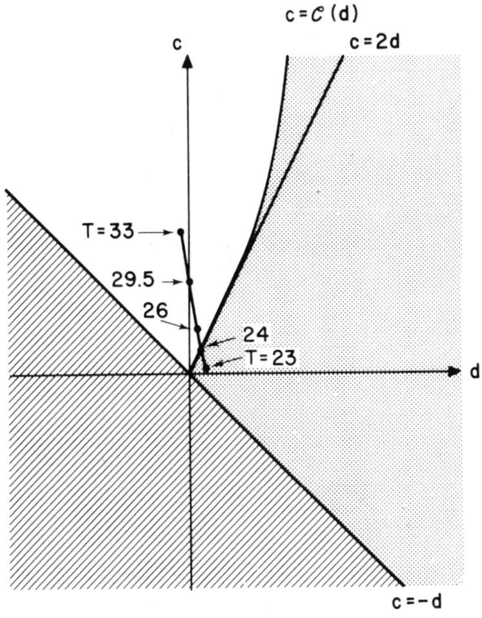

Fig. 12.6. A stability diagram in the d-c plane relevant to secular equation (29). The cross hatching designates the unattainable region while the shaded region corresponds to stability and the unshaded one to instability. The temperature curve appropriate for temperatures in interval I of Fig. 12.2 and with $\gamma = b_0 = 0.5$ is also included.

region, $c + d > 0$, the curve $c = C(d)$, which corresponds to $\lambda = i\theta$, $0 < \theta < 2\Pi$, in (29) and is given parametrically by

$$c(\theta) = \frac{(\sin\theta - \theta)\theta}{\cos\theta - 1}, \quad d(\theta) = 1 - \cos\theta + \frac{(\theta - \sin\theta)\sin\theta}{\cos\theta - 1}$$

separates the region of stability, $c < C(d)$ where $\text{Re}\lambda < 0$, from that of instability, $c > C(d)$ where $\text{Re}\lambda > 0$. Note that $\lim_{\theta\to 0} c(\theta) = \lim_{\theta\to 0} d(\theta) = 0$; hence $\lim_{d\to 0} C(d) = 0$. Also observe from Figure 12.6 that, for reasonably small values of d, the marginal stability curve, $c = C(d)$, can be approximated quite accurately by the line $c = 2d$ which is tangent to it at the origin in the d-c plane if one defines $C(0) = 0$.

What is of interest here for the purposes of biological control is that we determine the stability behavior of the community equilibrium point for relevant ranges of the governing parameters. Since McDaniel mite populations build to peak densities in mid to late summer, the analysis has been restricted to a period when the daily average temperature is typically within a 21-33 °C range. As mentioned earlier particular fixed values from that range have been taken for the temperature in this analysis. Motivated by our discussion of the previous section we have kept $b_0 = 0.5$ while γ has been assigned the same typical value of 0.5. Finally by performing a linear interpolation between the points $T = 21.1$, $D = 15$ and $T = 32.2$, $D = 16$ [6], we arrive at the following expression for D(T):

$$D(T) = 15 + \frac{T - 21.1}{11.1} \tag{30}$$

where D is measured in days and T, in this section, simply in °C. The behavior of the critical point under investigation can be determined for various temperatures by examining the curve included in the stability diagram of Figure 12.6 (see Table 12.1 as well) which has been plotted for T in the interval I from $T = 23$ where $r_1 = r_2$ to $T = 33$, the optimum temperature for r_1 (see Figure 12.2). Observe that for $23 \leq T < 24$ in that interval we have stability while for $24 < T \leq 33$ we have instability. The relevant features of Figure 12.6 are reproduced in Figure 12.7 which is a plot of σ_1 versus β. In that figure, $\beta = \gamma = 0.5$ corresponds to $c = -d$, while $\beta = \beta_c \cong 1.26$ corresponds to $c = 2d$ where β_c is that value of β, $.5 = \gamma < \beta < (2\gamma)/(1 - \gamma) = 2$, satisfying $\beta(\beta + 1)(\beta - \gamma) = 3[(\gamma - 1)\beta + 2\gamma] \cdot [\beta + 1/(1 + b_0)]$ generally or $\beta(\beta + 1)(2\beta - 1) = 3(2 - \beta)(\beta + 2/3)$ in our

TABLE 12.1

Values of the Parameter Pairs (d, c) of Fig. 12.6 and (β, σ_1) of Fig. 12.7 for Representative Temperatures from the Integral I of Fig. 12.2

T	σ_1	β	c	d
23.00	2.58	1.05	0.26	0.57
24.00	3.14	1.26	0.41	0.82
26.00	3.59	1.47	1.33	0.29
29.50	5.02	2.00	2.83	0.00
33.00	6.28	2.44	4.18	-0.16

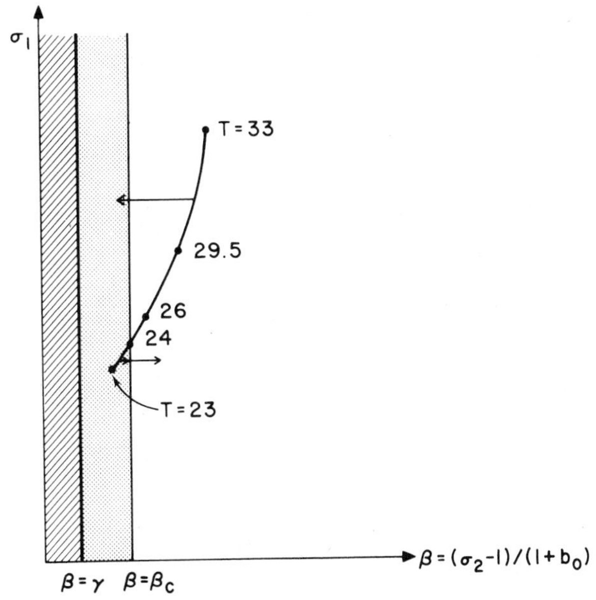

Fig. 12.7. A schematic plot in the β - σ_1 plane equivalent to Fig. 12.6. The regions in this figure are to be interpreted exactly as were the corresponding ones of Fig. 12.6. Again this has been drawn for $\gamma = b_0 = 0.5$ hence $\beta = 2(\sigma_2 - 1)/3$ and $\beta_c \cong 1.26$. The arrows indicate possible biological control strategies.

particular case. Observe that we have stability for $\gamma < \beta < \beta_c$ and instability for $\beta > \beta_c$. Since β is directly related to the intrinsic growth rate r_2 of the predator such a plot should be of aid to growers in planning control strategies (see below).

Before discussing the import of these results we shall consider the stability of the remaining two critical points of (18). Toward that end we transform the original equations of (12) into

$$\frac{dH}{d\tau} = \sigma_1 H(1 - H/K) - \delta HC/(H + b_1) \tag{31a}$$

and

$$C(\tau) = \int_{\tau-1}^{\tau} \left[\frac{\sigma_2 HC}{H + b_2}\right](s)ds \tag{31b}$$

where $H(\tau) = H(\tau D)$, $C(\tau) = C(\tau D)$, and $\delta = aD$. First from (31b) we observe that since $H/(H + b_2) < 1$, for $0 < \sigma_2 \leq 1$, $C(\tau)$ decreases at least exponentially to zero as $\tau \to \infty$. Using this fact in conjunction with (31a) it can be shown that the solution $H = K$, $C = 0$ is globally stable while $H = C = 0$ is unstable for $0 < \sigma_2 \leq 1$. We next perform a linear stability analysis of the critical points in question for $\sigma_2 > 1$ by considering a solution of (31) of the form

$$H(\tau) = H_0 + H_1(\tau), \quad C(\tau) = 0 + C_1(\tau) \tag{32}$$

Then substituting (32) into (31) and neglecting all nonlinear terms in the perturbation quantities we obtain the following equations for H_1 and C_1:

$$\frac{dH_1}{d\tau} = \sigma_1(1 - 2H_0/K)H_1 - \delta\left[\frac{H_0}{H_0 + b_2}\right]C_1 \tag{33a}$$

and

$$C_1(\tau) = \int_{\tau-1}^{\tau} \left[\frac{\sigma_2 H_0}{H_0 + b_2}\right] C_1(s)ds \tag{33b}$$

For the critical point (0, 0) setting $H_0 = 0$ in (33a) we obtain

$$\frac{dH_1}{d\tau} = \sigma_1 H_1$$

hence this point is locally unstable since $\sigma_1 > 0$ while for (K, 0) setting $H_0 = K$ we obtain

$$\frac{dH_1}{d\tau} = -\sigma_1 H_1 - \delta\left[\frac{K}{K + b_1}\right] C_1 \tag{34a}$$

and

$$C_1(\tau) = \int_{\tau-1}^{\tau} \left[\frac{\sigma_2 K}{K + b_2}\right] C_1(s)\,ds \tag{34b}$$

Since $\sigma_2 K/(K + b_2) < 1$ for $\sigma_2 < 1 + b_2/K$ while $\sigma_2 K/(K + b_2) > 1$ for $\sigma_2 > 1 + b_2/K$ from (34) we can conclude that this final critical point is locally stable for $1 < \sigma_2 < 1 + b_2/K$ and unstable for $\sigma_2 > 1 + b_2/K$.

We now summarize our stability results for all three critical points of (18) as follows:

1. For $\sigma_2 > 0$, the (0, 0) point is unstable.
2. For $0 < \sigma_2 < 1 + b_2/K$, the (K, 0) point is at least locally stable while for $\sigma_2 > 1 + b_2/K$ it is unstable.
3. For $0 < \sigma_2 \leq 1 + b_2/K$ there exists no community equilibrium point while for $\sigma_2 > 1 + b_2/K$ such a point exists, the possible stability of which is governed by the criteria graphically represented in Figures 12.6 and 12.7. Note that $\sigma_2 = 1 + b_2/K$ is a bifurcation point where the potential community equilibrium point (H_0, C_0), and the point (K, 0) coincide.

We now return to our analysis of that community equilibrium point and make some deductions about it which can be obtained upon examination of Figures 12.6 and 12.7. First note that in the limit of infinite McDaniel mite carrying capacity it can be shown from (20) and (29b) that

$$d \to \frac{-\sigma_1}{1 + \beta} \quad \text{and} \quad c + d \to \frac{\sigma_1(\sigma_2 - 1)}{\sigma_2}$$

Then even when $\sigma_2 > 1$ from Figure 12.6 one can see that now the community equilibrium is unstable since $d < 0$ and $c + d > 0$. Thus a necessary condition for a locally stable community equilibrium point is a finite such carrying capacity which is consistent with the simulation analysis of the previous section. Second from Figure 12.7 it can be seen that for relatively small values of $\beta < \gamma = 0.5$ no community equilibrium point can exist while for large values of this parameter $\beta > \beta_c \cong 1.26$ it is unstable. In the first instance the prey population goes to its carrying capacity in the absence of the carnivore while in the second that population goes to zero causing the system to crash due to overexploitation. Recalling that β is simply a measure of the productivity of the predacious mite species, these results are biologically reasonable.

Further observe that the temperature curve of Figure 12.7 can be used to deduce potential control strategies. For $23 \leq T < 24$ where the community equilibrium point is stable, if the population density of the prey, $H_0 = b_1/\beta$, is at an economically tolerable level then one does nothing. If for that interval this density is above such a level one can decrease it by continuously adding predacious mites and hence effectively increasing β. It may be necessary to increase β to such an extent that the resulting community equilibrium point becomes unstable and the system crashes. For $24 < T \leq 33$ the proper strategy might be to decrease β by *removing* predacious mites until a stable community equilibrium point is achieved which corresponds to a tolerable McDaniel mite density. These strategies are denoted by arrows in Figure 12.7.

Finally, it would be instructive to introduce temperature dependent initial population density functions in (16) that contained information on emergence from diapause [14] (a state of virtual suspended animation) in order to produce a model, in conjunction with (12), which, upon integration with the complete temperature-time profile, could provide quantitative agreement when compared with the observed data of Figure 12.4. Although the authors may perform such a "fine tuning" simulation at some later date, that analysis lies beyond the scope of this contribution.

REFERENCES

1. H. Von Foerster, "Some remarks on changing populations," *The Kinetics of Cellular Proliferation*, Grune and Stratton, New York (1959), 382-407.

2. J. W. Forrester, *Industrial Dynamics*, The M.I.T. Press, Cambridge, Mass (1971).

3. A. Hastings, "Spatial heterogeneity and the stability of predator-prey systems," *Theor. Popul. Biol.* 9(1977), 37-48.

4. C. S. Holling, "The functional response of predator to prey density and its role in mimicry and population regulation," *Mem. Ent. Soc. Can.* 45(1965), 1-60.

5. S. C. Hoyt, "Integrated chemical control of insects and biological control of mites on apples in Washington," *J. Econ. Entomol.* 62(1969), 74-86.

6. J. A. Logan, "Population model of the association of *Tetranychus mcdanieli* (Acarina: Tetrahychidae) with Metaseiulus occidentalis (Acarina: Phytoseiidae) in the apple ecosystem," Ph.D. Thesis, Washington State University, Pullman (1977).

7. J. A. Logan, D. J. Wollkind, S. C. Hoyt, and L. K. Tanigoshi, "An analytical model for description of temperature dependent rate phenomena in arthropods," *Environ. Entomol.* 5(1976), 1133-1140.

8. A. Oaten and W. M. Murdoch, "Functional response and stability in predator prey ecosystems," *Amer. Natur.* 109(1975), 289-298.

9. T. Royama, "A comparative study of models for predation and parasitism," *Res. Popul. Ecol.*, Supplement No. 1, The Society of Population Ecology, Kyoto, Japan (1971).

10. J. M. Smith, *Models in Ecology*, Cambridge University Press, Cambridge, England (1974).

11. L. K. Tanigoshi, S. C. Hoyt, R. W. Browne, and J. A. Logan, "Influence of temperature on population increase of *Tetranychus mcdanieli* (Acarina: Tetranychidae)," *Ann. Entomol. Soc. Am.* 68(1975), 972-978.

12. L. K. Tanigoshi, S. C. Hoyt, R. W. Browne, and J. A. Logan, "Influence of temperature on population increase of *Metaseiulus occidentalis* (Acarina: Phytoseiidae), *Ann. Entomol. Soc. Am.* 68(1975), 979-986.

13. R. L. Webster, "Mites of economic importance in the Pacific Northwest," *J. Econ. Entomol.* 41(1948), 677-687.

14. D. J. Wollkind, J. A. Logan and A. A. Berryman, "Asymptotic methods for modeling biological processes," *Res. Popul. Ecol.* 20(1978), 79-90.